Mechanochromism

David J. Fisher

Published by **Materials Research Forum LLC**
Millersville, PA 17551, USA

Published as part of the book series
Materials Research Foundations
Volume 52 (2019)
ISSN 2471-8890 (Print)
ISSN 2471-8904 (Online)

Print ISBN 978-1-64490-026-0
ePDF ISBN 978-1-64490-027-7

Distributed worldwide by

Materials Research Forum LLC
105 Springdale Lane
Millersville, PA 17551
USA
http://www.mrforum.com

Printed in the United States of America
10 9 8 7 6 5 4 3 2 1

Table of Contents

Materials Research Forum LLC
https://doi.org/10.21741/9781644900277

Introduction

Mechanochromism (sometimes referred to as piezochromism) is used here as an umbrella term for any sort of radiation-emission which occurs as a result of the stressing, deforming or breaking of solids. The most familiar phenomenon of this type, triboluminescence, has long been the basis of an amusing 'parlour trick' in which a 'boiled sweet' (a candy made from melted and re-solidified sugar) is squeezed in a darkened room by using a pair of pliers. This produces a distinct glow. The effect was indeed first studied[1] in candy-related crystals such as sucrose and tartaric acid. It was noted that it was most common in non-centrosymmetric (piezoelectric) substances – a correlation known as Zink's rule in the case of metal complexes having organic ligands - but was not observed in conductors. The process was therefore proposed to involve charge-separation during straining and fracture, thus generating an electric field. Breakdown of the atmosphere then excited N_2 molecules via electron bombardment, leading to N_2 photoluminescent emission and perhaps stimulated fluorescence. The N_2 line was often nicknamed the lightning-line due to its appearance in large-scale discharges from clouds.

There is certainly no shortage of candidate substances. An initial survey[2], made in the very earliest days of the present research effort, concluded that 36% of inorganic materials, 19% of organic materials, 37% of aromatic compounds and 70% of alkaloids are triboluminescent. In fact, almost half of all crystalline materials were thought to be triboluminescent.

In order to exploit fully the transducer possibilities, there has naturally been great interest in properly understanding all of the related atomic processes. Even the original researchers realized that materials such as quartz emitted light partly because they became hot. Further materials were discovered moreover which did not satisfy the criteria of the originally proposed mechanism. Triboluminescence sometimes coincided with the absence of centrosymmetry which was required for piezoelectricity, but only about half of triboluminescent sulfates were non-centrosymmetric. Many polymers were triboluminescent, and even some metals emitted radiation during fracture. Only those triboluminescent materials which are not photoluminescent are guaranteed to emit the signature N_2 'lightning' spectrum, and there remains little doubt that the triboluminescence emission is caused by electrical discharges. Visible light, electrons, ions and even radio signals have been detected during the fracture *in vacuo* of polymers, MgO crystals and sugar. The overall observations suggest that triboluminescence, electroluminescence and thermoluminescence can involve the same process; this being the migration of defects and the emission of light during the recombination of electrons with positive sites.

No single mechanism seems to explain all of the observed light-emission phenomena during stressing and fracture, although two distinctly different mechanisms have been proposed. Many materials exhibit a triboluminescence which involves discharge through the surrounding atmosphere, and the required electric field is typically generated by the piezoelectricity of the material. The discharge is characterized by N_2 emission lines in the triboluminescence spectrum, and by disappearance of the latter under liquids. Excitation of the surrounding gas could occur via electron bombardment, or excitation of the photoluminescence could occur via electron bombardment. Excitation of the photoluminescence could also occur via absorption of ultra-violet emissions from the excited gas. All three discharge mechanisms are moreover simultaneously possible. Triboluminescence usually involves a mixture of gas discharge emission, photoluminescence excited by gas emission and photoluminescence excited by electron bombardment. The mechanism which accounts for most of the light depends upon the material and the surrounding gas.

Some materials exhibit a triboluminescence which arises from processes occurring within the crystal, such as energy release due to the recombination of defects as they are driven into complementary sites by deformation or piezoelectric polarization. Shock waves can also generate triboluminescence in crystalline organic compounds, rather than simply destroying the compound and forming high-energy luminescent radicals.

In the case of metal surfaces, abrasion can cause the emission of electrons in a process which is known as the Kramer effect and which can be detected by measuring the voltage which is generated between the rubbing of two lubricated metallic surfaces. The resultant exo-electrons can induce chemical reaction, given that they can have kinetic energies of up to 4eV.

Interest in mechanochromism has exploded in recent years, and has advanced far beyond such novelty effects. In fact, triboluminescence *per se* will not be considered further here. The growing interest in mechanochromism, and a consequent burgeoning research effort, has been driven by the possibilities offered by the permanent rather than transient changes in colour which can be produced by stress. Permanent mechanochromic colour-changes are of immense practical interest for monitoring and recording stresses. There is, for instance, the possibility of turning such mechanochromic materials into useful optical pressure-sensors and structural damage sensors for the monitoring of stress, wear and fracture; sometimes referred to as being a 'self-reporting' capability. Mechanochromic and mechanoluminescent polymers, for example, could visually signal sub-micron damage and failure long before macroscopic cracks became detectable. One example would be the detection of incipient damage in composites, given that light emission occurs when a composite which contains a triboluminescent molecule is damaged. Monitoring the wavelength and magnitude of the emission then yields information on the degree of damage. Some pure organic crystals emit light during fracture. Pressure can produce changes in the photoluminescence spectra. External pressures can change the interaction between the ground and excited states of crystalline compounds and result in spectral effects. Because mechanochromic effects are visible to the naked eye as surface colour variations during mechanical deformation, mechanochromic photonic crystals can be used as a sensing device without requiring a power supply.

The rupture of covalent bonds requires relatively high forces, and so the dissociation of non-covalent interactions is increasingly popular for the preparation of mechanochromic materials. A protein-based force-nanosensor has been based upon two complexes which were variants of the Green fluorescent protein, and with the fluorescent spectra caused to form a donor/acceptor pair for fluorescence resonance-energy transfer. When the pair of complexes was held together by their covalent linking in the cavities of a third protein and connected to a cross-linked polymer, it accurately reported matrix strain changes by modulation of the fluorescence resonance-energy transfer signal. When one of the complexes was inserted into the interface between the glass fibers and the polymer matrix of a fibre-reinforced composite, fibre fracture or detachment led to unfolding of the protein and a resultant loss of fluorescence.

Force detection at sub-picoNewton resolution has been achieved by using a conjugated polymer with a semiconducting backbone, and doped with acceptor molecules along the chain. Upon extending the chain from the coiled to the elongated conformation, the local monomer density was reduced and the fluorescence emission spectrum appreciably shifted.

There are also more mundane applications: because even slight damage can cause a colour change, one can envisage the production of writing paper which does not require any writing implement other than a plain stylus. This recalls the products which resulted from the invention of micro-encapsulation some decades ago. Indeed some of the ideas which have been put under the mechanochromism umbrella-term resemble micro-encapsulation in that it has been suggested that excessive stresses could be signalled by incorporating dye-capsules that would be torn open by such stresses. Until recently however, the necessary aggregate and stress-transfer combinations had been limited to certain dye and polymer choices. The approach can now be extended[3] by tethering an excimer-forming cyano-substituted oligo(p-phenylene vinylene) fluorophore to the two ends of a telechelic poly(ethylene-co-butylene), and blending 0.1 to 2wt% of the resultant aggregachromic macromolecule into polymer matrices such as poly(ε-caprolactone), poly(isoprene), or poly(styrene-b-butadiene-b-styrene). All of these blends exhibit mechanofluorochromic responses and the monomer/excimer emission-intensity ratio can be used to relate the luminescence signal to the degree of deformation and to relaxation processes. This is nevertheless a relatively crude concept however and will not be considered further here.

Another relatively unsophisticated class of mechanochromic materials involves the incorporation of isotropically arranged monodisperse SiO_2 nanoparticles into a supramolecular elastomeric matrix. Such so-called photonic elastomers exhibit angle-independent structural colors, and the Young's modulus and elongation-to-failure of the as-formed materials can attain 0.24MPa and 150%. Their great elasticity leads to chameleon-like mechanochromic behaviour and, again like a skin, they are capable of mending scratches and cuts. In some cases luminescence-quenching material is coated with silica nanoparticles and this structure is in turn surrounded by a spray-dried raspberry-like dusting of nanostructured silica particles which host luminophores. During shearing, quenching and luminophore components which were spatially separated in the original supraparticles come into contact. This then causes irreversible quenching of the luminescence, and differing choices of the components can control the response-time and sensitivity, so that the resultant sensors can be sensitive and rapid in response, or monitor the effect of shear stress over a long time-period. Multi-colour mechanochromic polymer/silica composites can be created by using two different mechanochromophores. For instance, composites have been created having diarylbibenzofuranone in the silica-rich parts and naphthopyran in the polymer-rich parts. The composite then displayed blue, green or orange colorations according to the intensity of the mechanical stress. Colloidal crystals and glasses can also exhibit photonic effects: colloidal crystals have a high narrowband reflectivity while colloidal glasses have a low broadband reflectivity. A compromise is achieved simply by using two different silica-particle sizes with a repulsive interparticle potential. Monodisperse silica particles with a repulsive potential spontaneously take up a crystalline structure at volume fractions which are far below 0.74. When two different particle-sizes coexist, their arrangement is greatly affected by the size ratio and the mixing ratio. When the former is small, long-range order is partially preserved; resulting in a marked reflectance peak and a brilliant structural colour. When the size ratio is large the long-range order is rapidly reduced, together with the mixing ratio, although short-range order is preserved; producing a low reflectivity over a broad wavelength range and the appearance of faint structural colours. In recent work[4], silica nanoparticles, which served to trigger a colour change, were embedded in an elastomer so

Materials Research Forum LLC

https://doi.org/10.21741/9781644900277

as to form a bi-layer hybrid film. Upon stretching under ambient conditions, the hybrid film could change in both colour and transparency. It also exhibited excellent reversibility and reproducibility.

The mechanochromic materials to be discussed here can go far beyond offering just a simple warning colour-change, which might still have to be checked against a chart, as do pH-indicating chemicals. One can easily envisage a colorimetric elastomer being patterned so that a direct message such as 'HALT', for example, would appear when a critical stress was reached and thus deter continuation of the procedure.

The more subtle colour-changing phenomena currently under exploration offer the possibility of paper-like materials which not only can be written upon without a pen or pencil, but also more intriguing capabilities such as being able to erase that writing by, for instance, heating. Another choice of mechanochromic chemical might instead make the writing indelible. Stickers can be envisaged which would record any attempt at tampering. There is even the possibility of writing clandestine messages which could be revealed only by applying some suitable treatment. The opportunities offered by mechanochromic materials do indeed seem endless.

The number of suitable materials which have been discovered has also been rapidly increasing until they now make up a veritable menagerie. They are essentially all organic chemicals, and it is quite difficult to arrange them into a convenient catalogue because of their often complicated structures and lengthy chemical names. Because of the newness of the field, no general rules have yet evolved for predicting which materials will be mechanochromic, and which ones will not. One molecule may be so, while essentially the same molecule but with a slightly different configuration may not. One very noticeable trend however is for the colour-change to occur because of a change from a crystalline to an amorphous structure. Such changes can moreover be brought about even by mild grinding in a mortar and pestle or even mere smearing with a palette-knife.

Other phenomena can however account for colour-changes. Dimers in the bis(p-dimethyl aminophenylphthaloyl)ethane series, for instance, exhibit[5] mechanochromic properties due to the formation of free radicals during mechanical disturbance, and diperchlorates such as 2,2'-bis[2-(4-dimethylaminophenyl)-5-trimethylammonio-1,3-indandione] and 2,2'-bis[2-(4-dimethylaminophenyl)   -5-(2,4,6-triphenylpyridinio)   -1,3-indandione] possess[6] clear mechanochromic properties due to a reversible dissociation into free radicals.

Mechanically induced bond-breaking can create coloured forms of organic molecules such as spiropyran, a material which will be considered in some detail below. The latter molecules have a weak bond, between the nodal carbon atom and ethereal oxygen, which is easily broken and results in a reversible colour change from yellow to blue. The same colour change is engendered by grinding, which also causes a modest temperature increase of up to 6K. This mechanochromic effect is enhanced at liquid nitrogen temperatures. Essentially simultaneous intramolecular electron transfer and bond disruption are assumed to occur during mechanical stressing and to involve the formation of defects in the matrix: the molecules are excited around the defects and exhibit ring-opening, thus yielding a coloured molecule which is trapped in the crystal by the lattice forces.

Mechanochromism can also result from the reorganization of crystal packing. Some organics exhibit a colour-change from red to black when powder samples are ground, and the change is reversed by heating (280C, 2h) or by immersion in an organic solvent. It is theorized that the mechanical stress causes partial slipping of the molecules along stacking axes in the crystal, thus shortening the cross-stack interplanar distance. This generates a new band (750nm), and the black colour, due to excitonic interactions between transition dipoles following reorganization of the molecular arrangement during mechanical shearing. The reversal of the colour is then explained by relaxation of the disturbed lattice back into a stable state. During heating, the transition occurs due to lattice vibrations whereas immersion in an organic solvent loosens the lattice and permits the molecules to slide and rotate into a lower-energy state. It is not surprising therefore that spectral changes can result from mechanically-induced structural phase transitions.

Another important group of mechanochromic materials is that of dye-containing polymers. The thermoplastic polymers are made up of macromolecules; comprising a chain of atoms which are linked by single covalent bonds. Because the electrons which are associated with these bonds are in low-level orbitals, with a large energy-gap between the bonding and anti-bonding orbitals, such polymers are usually colourless. One exception is that of highly-conjugated polymers. The extended conjugation here reduces the energy-gap for some of the chain's binding electrons, and their mobility gives the polymer colour. The mobility can also impart semiconductive and electroconductive properties. The downside is the presence of chain rigidity and an adverse effect upon the viscoelastic behaviour.

Good thermoplastic polymer properties and a responsiveness to visible light can be combined by exploiting the principle that a coloured polymer can be created by dispersing a suitable dye throughout the originally colourless matrix. The macromolecules are thus structurally unaffected, but two-phase, unless the added dye is completely soluble in the host polymer. Interactions between the macromolecule and the chromophore govern the optical behaviour. Chromogenic materials react to thermal, mechanical, electrical or chemical stimulation by exhibiting changes in absorption, emission or reflective index. Chromophores with delocalised electrons can be made the basis of a sensor because they impart a variation in the opto-electronic properties in response to a disturbance.

Another approach involves the covalent insertion of chromophoric units into the macromolecular chain. This permits the chain to remain flexible, and the modified macromolecules constitute a polymer possessing a colour and intensity which is governed by the nature of the chromophore and its concentration. The resultant coloured material then has the structure of a co-polymer with a distribution of colourless and coloured monomers. Even naturally occurring coloured organic materials fall into one of the above two groups, with bio-macromolecules containing covalently-bonded dyes, dispersed natural dyes or protein platelets.

In mechanochromic materials which are made from dye-polymer blends, colour changes are associated with structural modifications of the arrangement of the dyes, dispersed in the thermoplastic polymer matrix, under mechanical solicitation. In binary systems where the dye is physically dispersed in the polymer, the optical response to mechanical inputs is tunable by controlling the interphase interactions and conformations. When a mechanical force is imposed on the polymer–dye system, macromolecule chain-slippage and reorganization cause disruption of the non-covalent interactions among the chromophore molecules and their dispersion in the matrix. Disruption of dye aggregates does not occur during elastic deformation of the thermoplastic, but only in the plastic range. The results of mechanically-produced dye-assembly aggregation and disaggregation are therefore irreversible. A comparison of the signal intensities of monomeric absorption and of the contribution arising from the aggregate form can be used to obtain information on polymer stretching and on mixing at the molecular level. Mechanical deformation of a film is able to unfold the macromolecular chains and create microfibrils in oriented crystalline and amorphous regions. This breaks up most dye aggregates, and their dispersion in the polymer matrix, and can produce marked changes in the emission properties and thus colour. The colour changes as a function of the monomer/excimer ratio change which is caused by film drawing.

Some materials exhibit so-called 'aggregation-induced emission' or 'aggregation-enhanced emission', where an off/on switching effect is produced by an external stimulus such as pressure. Such materials are more efficient emitters when the dye is in the aggregated state than when it is in the dissolved state. They offer potential application as light-emitting devices and chemosensors. The discovery of mechanochromic luminescence in aggregation-induced emission luminogens triggered a great deal of research on these previously rare materials. Aggregation-induced emission luminogens have twisted configurations which tend to produce loosely-packed structures that facilitate solid-state phase transformations. Amorphous films of aggregation-induced emission luminogens emit with an increased intensity under pressure, due to increased molecular interactions. Crystals of aggregation-induced emission luminogens on the other hand exhibit mechanochromic luminescence due to their amorphization by mechanical action. The most recent[7] example of aggregation-induced emission has been observed in a new molecular system, consisting of 3 positional isomers, which was prepared by attaching a methoxy group at various positions. These new isomers exhibited mechanochromic fluorescence with one of them, (E)-2-((((6-chlorobenzo[d]thiazol-2-yl)imino)methyl)-5-methoxyphenol exhibiting remarkable mechanochromic fluorescence in which the emission decreased in intensity, and red-shifted in wavelength, under mechanical solicitation. The emission-changes were due to a crystalline-amorphous transition and the previous state could be easily recovered by a recrystallization activated by solvent fumigation or immersion.

Mechanochromism has been observed in organometallic chromophores dispersed in polymethylacrylates. Grinding or rubbing could result in a marked red-shifted emission and other colour changes. The underlying mechanism involves a force-induced structural rearrangement which increases the number of shorter intermolecular interactions. It has recently been suggested[8] that multicolour mechanochromic polymer blends might be able to distinguish between stretching and grinding. That is, the blending of mechanochromic polymers with mechanochromophores, separately embedded in positions next to soft and hard domains, would produce a multicolour mechanochromism in response to distinct stimuli. In the case of dyes which are covalently linked to the polymer, mechanical stimulus can produce chemical transformation of the chromophore at the atomic level, and a resultant colour change. The responsible chromophore has to be positioned within the polymer chain in such a way as to suffer controlled mechanical perturbation. That chromophore must also have mechanically labile bonds that change via isomerisation or bond-scission.

Gold complexes, also to be considered in more detail below, can exhibit mechanochromic properties. Most complexes have a two-coordinated linear geometry, but T-form or trigonal-planar three-coordinated compounds and tetrahedral four-coordinated compounds are known. Three-coordinated complexes and their derivatives tend to exhibit properties such as luminescence, thermochromism, mechanochromism, vapochromism and solvatochromic luminescence.

When steric hindrance is not present, closed-shell gold centres are able to associate via aurophilic Au$\cdots$Au interactions, where the Au$\cdots$Au distances for intra-molecular or intermolecular aurophilic interactions range from 2.70 to 3.30Å and the strength of the attraction is 7 to 12kcal/mol. This attraction is thus comparable to that of hydrogen bonding. Aurophilic bonding permits the reversible self-assembly of supramolecular systems having various structures. The bonds are highly temperature-reversible and sensitive to physical and chemical effects. Luminescence is a prominent solid-state property of three-coordinated gold compounds at low temperatures. Aurophilic interactions have a marked effect upon absorption and emission energy levels, and these are very sensitive to external stimuli. The control of luminescence by such stimuli is pertinent to the construction of photo-active materials.

Some gold compounds change their luminescence colour in response to stimuli such as grinding or compression. Various types of mechanochemically induced solid-state transformation have been shown to lead to mechanochromic luminescence modification. The first evidence of mechanochromic luminescence in gold compounds was noted for a complex which not only exhibited luminescence thermochromism but also a marked mechanochromic luminescence. The linearly coordinated gold centres of various cations and anions were loosely connected, via aurophilic interactions, into infinite one-dimensional linear chains. Single crystals of the material were non-luminescent, but grinding produced strong green emissions. Phase transformations during grinding were ruled out as being the cause of the luminescence. The mechanochromic luminescence behaviour was instead attributed to grinding-induced mechanical stresses and bond rearrangement near to the surface; together with the formation of defect sites. Grinding generally increases the surface area and creates lattice defects.

Mechanochromic luminescence can also be produced by a crystalline-to-amorphous phase transition. In one non-emissive two-coordinated di-nuclear complex, the mechanochromic luminescence results from planarization of part of the ligand such as to increase the $\pi\cdots\pi$ conjugation of ring features. It will be found that a reversible mechanochromic luminescence can often be produced by crystalline-to-amorphous phase transitions, but only a few gold complexes exhibit that property. Reversion of complexes to their original state can occur in response to various chemical and physical stimuli, but understanding of the reversible luminescence switching of such complexes remains incomplete. It has been suggested that new emissions may arise from aurophilic Au$\cdots$Au interactions or from distortion of molecular conformations.

Mechanochromic luminescence can result from monocrystalline phase transitions. Such a mechanically-produced transformation can produce a marked luminescence colour-change. Under mild mechanical solicitation, the blue photoluminescence of monocrystals of one particular two-coordinated complex changes to yellow. The blue-emitting crystals are metastable polymorphs having a herringbone structure with C–H$\cdots\pi$ interactions, and longer aurophilic interactions between the molecules. The thermodynamically stable yellow-emitting polymorph features appreciable aurophilic interactions. A mechanical stimulus induces a single-crystal transition which self-propagates as the mechanical stimulus drives a state change throughout the entire material. The rearrangement under mechanical stimulation is attributed to attractive Au$\cdots$Au intermolecular interactions.

Mechanochromic luminescence and mechanochromism can even be promoted by chemical reaction. Some non-emissive three-coordinated di-nuclear complexes exhibit intense blue luminescence following mild crushing. Compounds having short Au$\cdots$Au bonds can be converted into a blue-emitting form by modest heating. The luminescence colour change is the result of molecular structure modification during solid-state chemical reaction. Di-nuclear molecules are connected by an infinite aurophilic helix via weak intermolecular Au$\cdots$Au interactions. Disruption of these weak intermolecular Au$\cdots$Au interactions, and rearrangement into aurophilically linked dimers can occur as a result of mechanical disturbance.

Phosphorescent materials have very long lifetimes as compared with those of fluorescent equivalents. In the case of organometallic complexes, the luminescence wavelengths depend markedly upon the ligand, the metal ion and the type of solid-state stacking. Grinding causes a difference in metal-metal interaction, with large spectral shifts. Many materials exhibit a switch in luminescence between two colour states under mechanical forces. The exact relationship between the emission wavelength and the magnitude of the force is still the subject of investigation. The occurrence of high-contrast colour changes following mechanical stimulation of course offers the possibility of practical application. In pure organic molecules, a high-contrast mechanochromic luminescence may occur upon changes in molecular structure or fluorescence–phosphorescence conversion.

Transient green mechanochromic behavior has been observed during the manipulation of certain epoxy resins and composites. This was not observed in other epoxy resins. It was proposed that the colouration was due to sulfenyl-radical formation of, aided by mechanically-assisted excision of aromatic disulfide cross-links. The colouration lifetime was closely related to polymer-chain mobility and the resultant likelihood of pairing such radicals. Sulfenyl radicals could survive for several hours at room-temperature before pairing-up to form new disulfide bridges; the polymer chains being effectively frozen. At higher temperatures, the polymer chains regained mobility and the green colouration disappeared within seconds. When the material was maintained at temperatures below -

Materials Research Forum LLC

https://doi.org/10.21741/9781644900277

20C, the colouration persisted for several days. Further investigation of this mechanochromism using computational chemistry suggested an explanation for why only para-substituted networks exhibited mechanochromism, and why the intensity of the transition for ortho-substituted network was very weak and greatly red-shifted.

Given the useful properties of gold complexes, but the obvious expense of that noble metal, there is considerable interest in finding cheaper analogues. Various types of externally stimulated luminescent responses have been observed for copper complexes. Many of the latter are based upon tetra-nuclear cubane-type Cu_4I_4 cores; a stable luminophore. Coordination-structure changes of the copper ion under external stimulation can moreover produce marked luminescence colour changes. One advantage of copper complexes, as compared with other metals, is the flexibility of the coordination structure which exists around the metallic center. That then permits easy changes in molecular structure under external stimuli, often leading to low luminescence quantum yields and short luminescence lifetimes. The mechanochemical synthesis of luminescent copper complexes can range from mononuclear species to coordination polymers comprising iodide-bridged copper-cluster cores. One possible cause of the mechanochromic behavior which is observed is suggested to be a slight shrinkage of the cubane-type core during grinding, and its effect upon the formation of the flexible tetra-nuclear core which is required for grinding-related luminescence.

The insertion of groups offering potential intermolecular non-covalent interactions is an effective route to developing mechanochromic materials, succinimide and trifluoromethyl groups being widely used. A simple mechanochromic polymorphic molecule based upon naphthyridine undergoes reversible molecular packing changes during grinding; resulting in reversible changes in fluorescence. The most common mechanochromic mechanism is related to changes in molecular packing or configuration. On the other hand, a simple herring-bone structure has been found to have three polymorphic forms, yielding red, orange and yellow emissions. The differing luminescences of red and orange crystals results from their differing molecular packings. The yellow and orange forms possessed the same crystal structure, but their different colours and emissions could be caused by differing particle sizes and surface structures. The molecules are linked via a tail-to-tail arrangement in some cases, and the intermolecular stacking between aromatic rings leads to the formation of molecular columns. These then stack together due to N-H$\cdots$O and C-H$\cdots$S interactions so as to form a three-dimensional structure. In other cases, the molecules are linked by N-H$\cdots$O and C-H$\cdots$O interactions so as to form a three-dimensional structure. A head-to-tail packing was observed in the orange crystals. Crushing one material could change the initial orange fluorescence to yellow. Other evidence showed that crushing merely broke the surface structure, without affecting the packing. This is a rare instance of a mechanochromic molecule which does not exhibit changes in molecular packing or conformation.

Another interesting group of materials is that of 1,2-dioxetane derivatives, where load-accelerated cleavage generates chemiluminescence. Luminescence is to be preferred over fluorescence because the signal is transient rather than additive. This yields a higher contrast, and greater spatial and temporal resolution of bond-breaking, but the emission is very weak. Dioxetane is therefore often coupled to an acceptor fluorescent molecule in order to increase the quantum yield.

Mechanochromic polymers exhibit colour generation or change when they are mechanically loaded along the polymer chains. In early studies, spiropyran – the most widely studied mechanophore - was introduced covalently into the macromolecular backbone; either as the center of linear elastic polymethacrylate or as a cross-linker of glassy polymethylmethacrylate. During deformation, the 6-π electrocyclic ring-opening of spiropyran into red merocyanine occurred and the colour of a polymer film changed from colourless to red. Measurement of changes in the absorbance of the polymer permitted visual C-O bond scission and quantitative stress detection. Polymeric spiropyran could be covalently linked to polycaprolactone, polyurethane, block copolymers and hydrogen-bonded supramolecular polymers. Two general types emerged: ring-opening or radical-forming. Naphthopyran, for example, exhibited an orange-yellow colour arising from ring-opening. Mechanochromic behavior occurred only when substitutions were made at the 5-position. The resultant polymers were mechanically inactive when chains were attached at the 8 or 9 positions of the naphthopyran group. This trend was captured by density functional theory predictions, and it was attributed to differences in the alignment of the target C-O pyran bond with the direction of the applied force. Rhodamine has also been used as an effective colour-producing mechanophore in a polyurethane-based elastomer. Calculations suggested that the C-N bond of the rhodamine was quite weak and that the minimum energy required for force-induced ring-opening was as low as 32kcal/mol. The rhodamine was covalently joined to the cross-linked polyurethane network, and colour and fluorescence changes which occurred under pressure originated from the chemical transformation of rhodamine from a twisted spirolactam into a planarized zwitterionic structure. In the case of another rhodamine-based mechanochromic elastomer, the stressed elastomer underwent a fluorescent colour change from pale-blue to red. A simultaneous transformation from colourless to dark-red was observed, and removal of the stress led to a yellow fluorescence. Polymer mechanochromism also occurred in radical-forming systems. In one case, di-arylbibenzofuranone was used as a colour-changing mechanophore. This could act as a covalent cross-linker in polyurethane. Because of its special properties, the radical system permitted the direct quantitative evaluation of polymer chain scission.

Having briefly described some of the major materials groups which have attracted interest with regard to their mechanochromic properties, it will be instructive to review them in more detail. It soon becomes evident that, in spite of the numbers and variety of the materials which have been studied, there is precious little known in the sense of a general theory for predicting whether a given substance will be mechanochromic and certainly not for predicting what colours would be involved. At the current stage of development of the field, one nevertheless has an extensive palette of colours available for potential exploitation.

Although most effort in the field of mechanochromic luminescence has been expended on single-component dyes which can change their solid-state emission between two colours, it is possible[9] to produce tricolour behaviour by mixing two organic dyes which exhibit poor, or no, mechanochromic luminescence. This is due to a transition, of the fraction possessing poor mechanochromic luminescence properties, from a crystalline to two amorphous states: perfectly amorphous and proto-crystalline amorphous.

Spiropyran

This is the most extensively studied mechanophore. It undergoes a reversible force-induced (a few hundred picoNewtons) ring-opening which transforms spiropyran into merocyanine; a transformation which is associated with a visible change from colourless spiropyran to violet merocyanine. Covalent integration of spiropyran into a polymer backbone was found to be a successful strategy for preparing printable force sensors. When inserted into a polymeric chain or used for cross-linking, it is a very useful tool for mapping stress distributions. In a stretched polymer which is phase-separated into soft and hard domains, the anisotropy of the fluorescence polarization yields information on the orientation and degree of alignment of the strained polymers of each phase. Some spiropyran derivatives have been characterized in terms of their force versus rate relationships. The quantitative assessment of local stresses depends upon the environment, and upon the deformation-mode and speed. If the relaxation processes of the material are more rapid than the molecular force response, estimation of the spiropyran activation time is more difficult. Mechanical activation of spiropyran occurred during creep tests at stress levels which were below the yield stress of the polymer matrix. Upon increasing the imposed stress, the time required for the onset of spiropyran activation decreased. The onset of activation corresponded to a maximum in the strain rate, suggesting that mechanochemical activity requires a stress-enhanced chain mobility and large-scale polymer rearrangement. Activation of spiropyran-type molecules tends to occur only at large macroscopic strains, and to correspond to irreversible plastic deformation. Suitable design can lead to fluorescence activation at about 14% strain. Three-dimensional printing permits the creation of complex macroscopic structures having spatially localized spiropyran-containing regions. Tensile straining of the printed structures transformed the spiropyran into merocyanine and thereby allowed visual identification of regions suffering differing degrees of stress. The onset of activation of each region of spiropyran was sequential during the elongation. A more detailed quantitative relationship still needs to be established between the merocyanine intensity (content) and the applied force in order to be able to predict failures on the basis of the mechanochromic response.

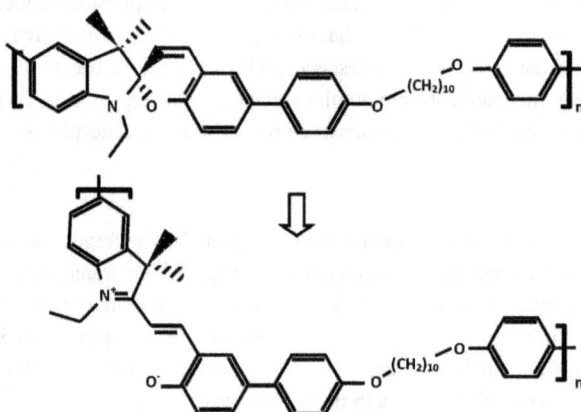

Figure 1. Conversion of spiropyran (colourless)
to merocyanine (coloured) by shear forces

A spiropyran-based mechanochromic co-polymer was used[10] in the microwave-assisted synthesis of a mechanochromic alternating co-polymer, having a molecular weight of up to about 174kg/mol, via Suzuki-Miyaura polycondensation. Dehalogenation, oxidative deborylation and spiropyran cleavage were the limiting factors that impeded any further increase in molar mass. Side-reactions such as protiodeborylation were not observed. The embossing of films yielded the coloured merocyanine (figure 1) co-polymer, which underwent thermally-aided back-reaction. It was proposed that the barrier to spiropyran

→ merocyanine conversion involved two factors: one being related to the colour change, and the other to internal bond-reorganization. The barrier height was 1.5eV; thus suggesting that the easy thermally-aided back-reaction was due either to residual energy stored in the deformed polymer matrix, or to a merocyanine isomer that was not in the most stable thermodynamic state.

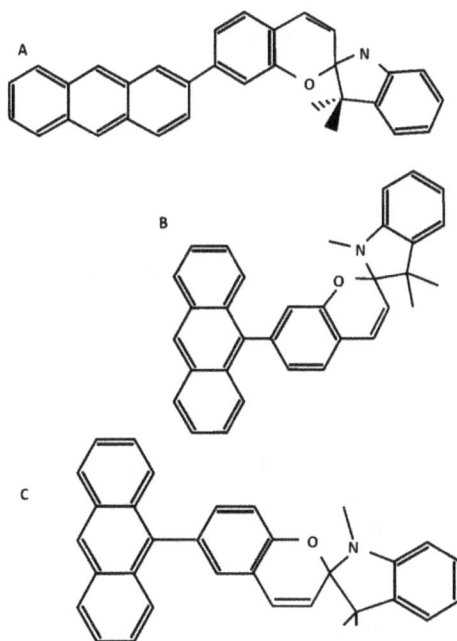

Figure 2. Structures of spiropyran-anthracene molecules

It is well-known that the electronic transport in zig-zag graphene nanoribbons becomes spin-polarized upon applying an electric field across the nanoribbon width, but the required fields are large and difficult to apply. It has been suggested[11] that a spiropyran-based polymer which was non-covalently deposited onto such a nanoribbon could act as a dual opto-mechanical switch and modulate its own spin-polarization. Calculations showed that, under mechanical stress or photo-absorption, the spiropyran could isomerize from a closed-configuration ground-state into a zwitter-ionic excited state. This would then result in a large change in dipole moment and alter the electrostatic environment of the nanoribbon. The electronic spin-distribution in nanoribbon-spiropyran hybrid material could then be reversibly modulated using non-invasive optical or mechanical means and obviate any need for high external electric fields. In particular, a mechanical stress giving rise to 39% elongation or photo-absorption at between 250 and 450nm could create a large dipole field and generate a potential energy difference, between nanoribbon edges, and destroy spin symmetry. Even when using unmodified spiropyran material, it was possible to produce an almost 50% change in band-gap, between up-spin and down-spin states, over the nanoribbon substrate.

A photochromic polymer which exhibits mechanochromic behavior has been prepared[12] via the ring-opening polymerization of ε-caprolactone, using difunctional indolino-spiropyran as an initiator. The presence of such a photochromic-initiating species in the polymer chain led to a mechanochromic effect, in which deformation of polymer films led to ring-opening of the spiro C-O bond to form the familiar coloured component, merocyanine. Irradiation with ultra-violet light produced a negative change, in the polymer stress, which was of the order of 50kPa. A sharp increase in the relative molecular energy and absorption wavelength, together with a marked decrease in the spiro-oxygen atom charge, occurred at a molecular elongation of more than 39%. Three spiropyran-containing poly(ε-caprolactone) polymers were later[13] filamentized and used to print single- and multi-component tensile testing specimens three-dimensionally. Filament production and printing did not degrade the spiropyran units or polymer chains, and the mechanical properties were as expected. Also prepared were poly(ε-caprolactone) which contained a spiropyran unit that was selectively activated by mechanical action. Multi-component specimens which contained two different spiropyrans permitted the selective activation of various parts of the specimen, depending upon the applied stimulus. Three-dimension printing can thus create, for example, a force sensor that permits the assessment of a load by means of direct visual observation.

The fluorescent mechanochromism of three synthesized anthracene designer molecules (figure 2) was studied in the original crystalline state and in monocrystalline form[14]. The molecules had small structural dissimilarities but exhibited very different responses to anisotropic grinding and isotropic hydrostatic pressure in both sample states. Manual grinding converted originally ordered crystallites into an amorphous form but this only slightly shifted the emission to longer wavelengths. This was taken to imply that ring-opening isomers were rare in this system. Hydrostatic pressure meanwhile forced together the anthracene units of single crystals, leading to increased π-π interaction and the isomerization of the spiropyran component into a flatter and more conjugated merocyanine form. This produced a gradually red-shifted emission, and red fluorescence.

Single crystals of two of the materials exhibited a high-contrast multi-colour change and the largest-yet observed emission red-shift: >190nm. There is thus a marked dependence of mechanochromic properties upon the position of anthracene-anthracene and anthracene-spiropyran frameworks.

In situ micrographs were used to observe the color-changes of single crystals. Single crystals of material-A were colourless and transparent under normal pressure and visible light. A red colouration appeared at high pressures. The single crystals exhibited a deep-blue emission at 439nm (small shoulder at 470nm). Emissions like these depended upon the anthracene position. The colours of anthracene crystals were related to the degree of π-π interaction between adjacent units. Colour changes involving a red-shift implied increased π-π overlapping between the anthracene units. The predominant emission of material-A, at 439nm, corresponded to neighboring anthracene rings having weak π-π interactions. When the pressure was increased from 3.12 to 6.27GPa, the corresponding emission red-shifted continuously from 509 to 552nm and the colours varied from bright-blue to green. This suggested that the π-π interaction between adjacent anthracene units increased so as to give a symmetrical stack having a sandwich-like form. When the pressure was increased to 11.32GPa, the emission band was centered at 633nm and was attributed to pressure-induced isomerization of spiropyran from a ring-closed state to a ring-opened merocyanine form with a plane structure. Pressure-induced molecular planarization generally increased the oscillator strength and led to a fluorescent intensity decrease at high pressures. When the pressure was returned to ambient, the high-pressure red colour returned to green but not to deep-blue; perhaps due to damage caused to the stacking of anthracene units during compression.

Material-B originally exhibited a broad emission band centered at 492nm, giving a bright-blue colour. Under increasing pressure, the colours varied from cyan to green, yellow and then orange with the emissions changing from 492 to 606nm at 9.99GPa. This was attributed to a closer packing of the anthracene rings, with intense π-π interaction and ring-opening of the spiropyran due to compression. With further increases in pressure, the emission intensity markedly decreased; reflecting an increase in excitation transfer.

The isomerization of spiropyran in single crystals of material-C occurred at higher pressures. The blue colour became green, and the emission continuously red-shifted from 422nm under ambient pressure to 457, 476 and 506nm at 3.11, 6.01 and 9.02GPa, respectively. A green colour appeared, and the wavelength shifted to 544nm, when the pressure was increased to 12.20GPa. No obvious red colour appeared, but a weak fluorescence peaking at 589nm was detected at 15.20GPa.

The UV-Vis spectra of single crystals of all of the materials changed slightly at low pressures. When the pressure was increased to 11.32, 9.99 and 14.20GPa, materials A, B and C, respectively, new absorption peaks appeared and gradually increased under further compression; indicating isomerization of the spiropyran components due to high pressures. The overall results indicated that hydrostatic pressures forced the anthracene units to approach each other, resulting in red-shifted emissions, and that compression increased π-π interaction to a higher level, as compared with grinding. The molecular structure and the method of applying force, such as anisotropic grinding and isotropic pressuring, had a marked effect upon the mechanochromism. Thus no red fluorescence was detected from material-A crystallites at atmosphere pressure in spite of vigorous grinding, but was observed under high pressures. This suggests that compression distorts crystal structures so as to facilitate the ring-opening reaction of spiropyran. Higher pressures trigger a red fluorescence which arises mainly from the merocyanine form of spiropyran, together with the enhancement of π-π interaction between anthracene units. The isomerization of spiropyran required more than 10GPa of pressure.

A rare example of a luminescent material exhibiting dual mechanochromic and photochromic responses was a synthetic molecule which used a dipeptide as a spacer to link rhodamine-B and spiropyran components[15]. This molecule possessed efficient photochromic properties in solution and in the solid state. High-contrast independent fluorescence switching also occurred under the influence of external forces. A two-step ring-opening reaction and subsequent fluorescence resonance-energy transfer between donor-acceptor pairs within a single molecule imparted successive colour-switching capabilities under mechanical control.

The macroscopic behaviour of silicone elastomers, including cross-links made up of three spiropyran region-isomers, have been studied[16]. The spiropyran was connected to the network via an identical attachment-point on the indoline fragment, and regio-isomeric ortho, meta and para attachments were made to the spirocyclic C-O bond on the benzaldehyde portion. The relative colorimetric responses of the regio-isomers, under quasi-static uniaxial tensile loads, were: spiropyran$_{ortho}$ > spiropyran$_{meta}$ > spiropyran$_{para}$. This was consistent with the predicted mechanical sensitivity of the regio-isomers. The extrapolated onset-strains for detectable activation of the region-isomers were indistinguishable however and occurred at about 90% uniaxial strain. The ratiometric response of the isomers was constant over the investigated strain range of 90 to 135%.

A composite of pyrene and spiropyran has been produced via separate linking at the same peptide-based dendron[17]. Co-assemblies having various weight ratios were prepared in which the morphology could be tuned. The resultant stable organogel was doubly-switchable, in that it could be triggered by light or heat. Within the xerogel, the dendrons formed rigid rod nanofibers having large diameters and acted as a rigid skeleton through which an interwoven fibrous structure grew. The original xerogel powder not only exhibited sequential high-contrast three-coloured switching, from dark-blue to bright-cyan to red, under an external force but also multi-state photochromic properties. These behaviors were due mainly to transitions of various excimers of pyrene and to the induced ring-opening reaction of spiropyran.

Mechanochromic nanofibers, prepared by electrospinning, have been used to produce nanofiber-poly(dimethylsiloxane) composites having isotropic and anisotropic responses[18]. Due to chain alignment of spiropyran co-polymer chains within the nanofibers, only very small strains were required in order to yield a mechanochromic response. Composites having aligned isotropic nanofibers could exhibit anisotropic and isotropic mechanochromic behaviours. Due to the special substitution pattern of spiropyran in the co-polymer, the mechanochromic response of the composites underwent rapid reversibility upon force removal. The use of such highly sensitive mechanochromic nanofibers as fillers in composites would permit the detection of directional stresses and strains.

Composites can be produced in which stretchable electronic properties are coupled with a mechanochromic ability which produces visual signals. Such properties are obtained by covalently incorporating spiropyran into poly(dimethylsiloxane)[19]. The strain at which colouration appears can be controlled by layering silicones, having differing moduli, into a composite. The colour-onset can be tailored so as to indicate when a specified frequency of a stretchable antenna has been attained.

Poly(meta,meta,para-phenylene) polyarylene is tough, has a molar mass of 60 to 96kg/mol, a modulus of 0.9GPa and a fracture elongation of 300%. Its co-polymers with para,para-spiropyran are mechanochromic[20]. Strained samples instantaneously lose their colour when a force is released, and this is attributed to the fact that the polyarylene matrix permits the build-up of sufficiently large forces to be transferred to the spiropyran, and to the relatively unstable merocyanine form. The latter, when covalently incorporated into the matrix has a half-life time of 3.1s, as compared with the 4.5h found for 6-nitro substituted spiropyran.

A self-reporting polythiourethane-tetrapodal-ZnO composite has been produced[21] by using spiropyran, at concentrations as low as 0.5wt%, as an additive. Heat, ultra-violet light and mechanical force caused the spiropyran to exhibit reversible isomerisation; as reflected by a reversible colour change. The spiropyran concentration was maintained at 0.5wt% while varying the tetrapodal-ZnO concentration from 0 to 7.5wt%. The latter acted as a prism, and its light-scattering effect created the illusion of a uniform colour distribution. The interconnected network of the tetrapodal-ZnO, embedded in the polythiourethane matrix, improved the mechanical stability of the polymer and led to an impact resistance of up to about 232kPa. Polythiourethane-spiropyran also proved to be a potential thermal sensing coating, due to its temperature-sensitivity. Because of the broad green luminescence band (circa 535nm) of tetrapodal-ZnO, the coloured merocyanine which absorbed in that range switched back to spiropyran at that wavelength. High concentrations of tetrapodal-ZnO reduced the effect of ultra-violet light. By exploiting this property of tetrapodal-ZnO, it is possible to produce a switchable system which reacts to several separate stimuli.

A multi-colour switching mechanochromic aggregation-induced emission material, having a fluorescence spectrum which spans almost the entire visible region, was obtained[22] by covalently attaching spiropyran to tetraphenylethylene. The covalent attachment permits tuning of the fluorescence of the spiropyran-tetraphenylethylene combination from zero colour-change, to two-colour switching and to three-colour switching by grinding. This is attributed to amorphization of the tetraphenylethylene and to isomerization of the spiropyran. Under hydrostatic pressure, multi-colour switching over a wide spectral range was obtained, due to the existence of distinct pressure-thresholds for supramolecular and chemical structural changes. The spiropyran enhanced the response.

Force-sensitive acrylic latex coatings can be created by covalently incorporating spiropyran. The latex was obtained via the emulsion co-polymerization of butyl acrylate, methylmethacrylate and vinyltriethoxysilane[23]. The condensation of hydrolyzed vinyltriethoxysilane provided interparticle cross-linking and improved the mechanical properties of the coatings, and mechano-activation of the spiropyran-containing latex was noted. Increasing the content of the spiropyran intra cross-linker led to a higher stress sensitivity and to a lowering of the critical stress required for mechano-activation. Increasing the content of the vinyltriethoxysilane interparticle cross-linker led to a higher critical stress for spiropyran mechano-activation but hardly affected the stress sensitivity. The temperature had an appreciable effect upon mechano-activation. A lower strain-rate encouraged greater spiropyran-to-merocyanine conversion.

A high-contrast tri-state fluorescent switch which exhibits both an emission colour change and on/off switching can be created[24] in a single molecular system by combining aggregation-induced emissive tetraphenylethene with spiropyran. Unlike most solid-state fluorescent switches, this material exists only in the amorphous phase, in the ring-closed form, due to its very asymmetrical molecular geometry and weak intermolecular interactions. This leads to a grinding-inert cyan emission in the solid state. The amorphous phase induces a structural transition from the ring-closed to ring-open form, together with the fluorescence off state. Structural transition leads to a planar conformation and high dipole moment, together with strong intermolecular interactions and good crystallinity. A reversible choice between any two of the three (cyan/orange/dark) possible states can be made via a mechanical-force/solvent treatment. Such a fluorescent switch is highly promising for use in mechanical sensing.

The complexing of rhodamine or spiropyran with cyclodextrin has been investigated. This imparted a perfect mechanochromism to various fluorophores by introducing a supramolecular system. The non-covalent chemical modification and easy preparation made this approach potentially useful for practical application. The strong hydrogen bonds provided by the cyclodextrin were crucial in triggering the mechanochromic conversion. That is, the hydrogen bonds take hold of both of the weak chemical bonds of the fluorophore and firmly confine the fluorophore to the cyclodextrin cavity[25]. The hydrogen bonds also promote the aggregation of complex inclusions into large ordered arrays and strengthen molecular interaction. The weak chemical bonds then concentrate greater external force and stretch more readily during shearing.

The behaviour of a spiropyran and pyrene derivative under isotropic compression in a diamond anvil cell involved a continuous mechanochromic process, with a colour-change from blue to red. Tri-colour switching could be obtained by means of sequential, slight and heavy, grinding[26]. The mechanochromic response could be adjusted by controlling the types and strengths of mechanical loading. This suggested a simple strategy for the design of multi-colour switching molecules and a control of mechanochromism under pressure on the basis of supramolecular changes and molecular structural changes.

When isomerization of spiropyran in crystals was carried out under high pressure, the mechanochromic response was observable using the unaided eye. The equilibrium constant increased with increasing pressure, and it was suggested[27] that the negative volume of reaction governed isomerization under high pressures.

The high strain-rate response of spiropyran in polymethylmethacrylate, a system in which reversible bond-scission in spiropyran under local tensile forces converts it to fluorescent merocyanine (figure 3), was studied by using a Hopkinson-bar to apply compressive loads at rates of 10^2 to 10^4/s. This resulted in a marked activation of the spiropyran near to fracture surfaces, although much of the observed activation was attributed to thermal effects arising from the high-rate fracture. Fluorescence-mapping of the fracture surfaces suggested that mechanical activation may occur in craze-like regions during fibril rupture[28]. The thermal response of spiropyran is thus itself useful for monitoring the plastic heating of regions during dynamic fracture. In fact, a complication arises from the energy-input which occurs during confocal imaging measurements. This can itself affect activation of the mechanochromic species. A study was made[29] of laser-based imaging of spiropyran which was integrated into polymethylacrylate and polymethylmethacrylate. Localized stresses and temperatures were monitored during high-rate compressive loading and dynamic fracture. The laser illumination of spiropyran in polymethylacrylate indicated that there was a strong excitation-wavelength and power dependence.

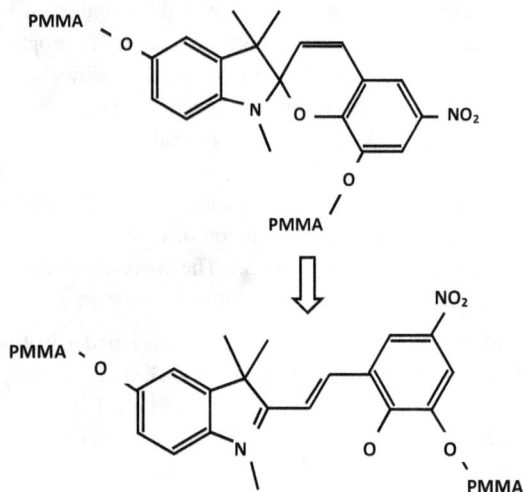

Figure 3. Effect of stress on spiropyran-polymethylmethacrylate

Hydrogen-bonded 2-ureido-4-pyrimidinone and covalent spiropyran have been incorporated into a single polymer structure so as to tailor their mechanical and mechanochemical responses[30]. The spiropyran was incorporated into the backbone of a pre-polymer which was also end-capped with ureidopyrimidinone or urethane. Strong mechanochromic reaction of the spiropyran occurred in bulk films of ureidopyrimidinone-containing materials while much weaker activation occurred in urethane-containing forms, as reflected by their stress-strain responses.

Differences in the magnitudes of the supramolecular interactions led to various degrees of chain orientation and strain-induced crystallization in the bulk and their ability to transfer loads to the mechanophore[31]. Tensile testing confirmed the achievement of a high strength, high fracture elongation and high toughness, due to successive fragmentation of hard domains made up of stacks of ureidopyrimidinone dimers and to the dissociation of those dimers. The fragmentation and dissociation occurred before mechanical activation of the spiropyran. The stress which presaged catastrophic failure was signalled by a colour change which coincided with strain-hardening. In a similar spirit, mechanical activation of spiropyran in a doubly cross-linked polyurethane elastomer was studied. As well as chemical cross-linking, the material involved polytetrahydrofuran as soft segments and, again, the hydrogen-bonded 2-ureido-4-pyrimidone as hard segments. Because of the ring-opening conversion of spiropyran to merocyanine in the strained state, and isomerization about the methane bridge of merocyanine in the relaxed state, the material exhibited two colour-changes[32]. An increased tensile strain-rate led to a stiffer and stronger elastomer and to earlier activation of the spiropyran. The conversion-point of spiropyran to merocyanine coincided closely with strain-hardening in the stress-elongation curves, and the two spiropyran colour transitions could be used to monitor fracture of the elastomer during crack propagation.

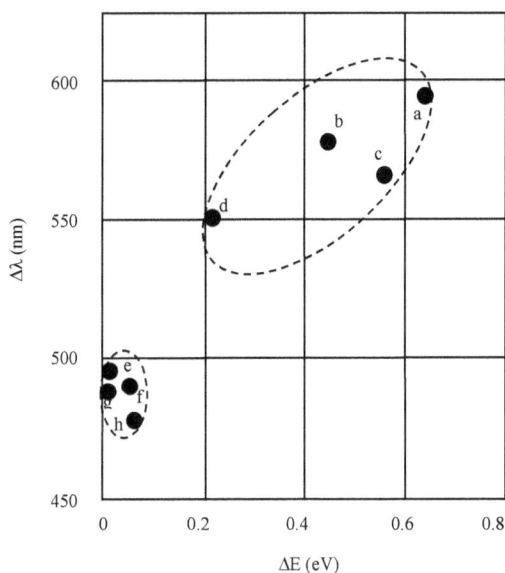

Figure 4. Emission wavelengths versus calculated energy difference between highest occupied molecular orbitals for various gold complexes: a) PF_6, b) NO_3-3MeOH , c) NO_3-2EtOH, d) $B(C_6H_4F-4)_4$, e) I, f) Br, g) NO_3, h) Cl

In the latest research[33], a spiropyran having 3 attachment points was compared with two already-popular spiropyrans, having just 2 attachment points in the same matrix, in order to find out whether geometrical or electronic effects governed the force-response. The attachment points on the chromene ring of the 3 spiropyrans were identical, while there were various attachment points on the indoline ring. Colour-analysis and absorption measurements showed that, when changing the attachment points of spiropyrans, both electronic and geometrical effects affected the mechanochromism, but geometry played the major role.

Gold Complexes

Tetrahedral gold complexes containing a diphosphane ligand, [Au(1,2-bis(diphenylphosphino)benzene)$_2$]X, where X was Cl, Br, I, NO$_3$, BF$_4$, PF$_6$ or another adduct, have been prepared[34]. The complexes fell naturally into two categories (figure 4): in one category, the complexes contained relatively small counter-anions and two 1,2-bis(diphenylphosphino) benzene ligands were symmetrically linked to the central gold atom; leading to an intense blue phosphorescence. In the other category, the complexes had large counter-anions and two 1,2-bis(diphenylphosphino)benzene ligands were asymmetrically linked to the gold atom; leading to a yellow or yellow-orange phosphorescence. The difference in phosphorescence colour between the two categories of complex was attributed to the change in the structure of the cationic component of the complex. The symmetry-reduction caused by the large counter-anion of a complex in the second category led to destabilization of certain levels, leading to a red-shift of the emission peak. Symmetry-reduction was also responsible for phosphorescence-colour alterations caused by mechanical grinding.

Non-covalent interactions are often highly sensitive to physical or chemical stimuli, which trigger alterations in the molecular and crystal structures and hence lead to changes in the macroscopic properties of gold compounds. Some gold complexes exhibit useful stimulus-responsive properties such as mechanochromism[35].

The lifetimes of the transient phases of luminescent mechanochromic gold isocyanide complexes could be tuned[36] via modification of the ends of flexible side chains. Upon grinding, these complexes exhibited an initial emission colour change from blue to yellow. The yellow emission spontaneously changed to blue or green, but with various lifetimes of the yellow-emitting transient phases; depending upon the nature of the substituent.

Materials Research Forum LLC
https://doi.org/10.21741/9781644900277

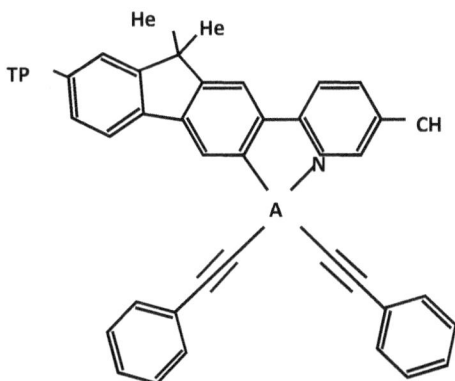

Figure 5. Structure of bis(alkynyl)gold(III) N^C complex A

Donor-acceptor luminescent bis(alkynyl)gold N^C complexes have been synthesized[37] which exhibited photoluminescence quantum yields of up to 0.81, together with a reversible mechanochromic luminescence. This mechanochromic luminescence could be easily tuned by introducing suitable substituents into the pyridine ring (figures 5 to 7). As a result of grinding, there was a marked luminescence colour-change from green to red in solid samples (table 1).

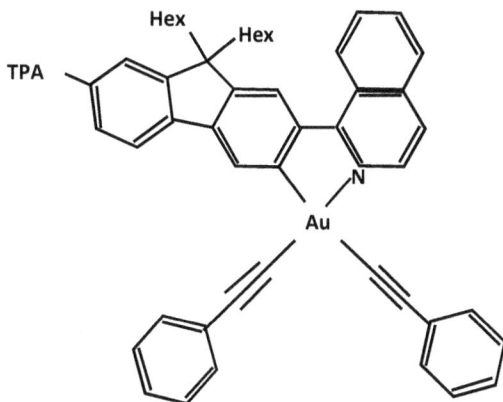

Figure 6. Structure of bis(alkynyl)gold(III) N^C complex B

The coordination-driven self-assembly of block co-polymers with gold complexes can lead to the formation of spherical micelles which exhibit a marked increase in phosphorescence. The luminescence level increased unexpectedly[38] upon increasing the degree of polymerization of the non-coordinated block. One of the gold-containing metallopolymers exhibited a reversible mechanochromic luminescence.

Forty-eight substituted isocyanide gold complexes were examined, and twenty-eight of them were found to be mechanochromic[39]. Two compounds exhibited a crystalline-to-crystalline phase transition during mechanical treatment. Monocrystals of as-prepared and ground forms of a new CF_3-CN complex remained, and theoretical calculations indicated that a mechanically induced red-shifted emission of the CF_3-CN was caused by aurophilic interactions. A comparison of the crystal structure of CF_3-CN with those of other complexes indicated that weaker intermolecular interactions in the as-prepared form were an important factor in mechanically induced crystalline-to-crystalline phase transitions.

Table 1. Luminescence colours of bis(alkynyl)gold N^C complexes

Complex	Condition	λ(nm)
A	as-received	504
A	ground	553
A	recovered	502
B	as-received	555
B	ground	630
B	recovered	549
C	as-received	554
C	ground	640
C	recovered	542
D	as-received	562
D	ground	610
D	recovered	567

Interconvertibility of the tetra-coloured solid-state photoluminescence of a gold isocyanide complex (figure 8) under various external stimuli was studied[40]. Soaking the complex in acetone produced a blue emission. Subsequent removal of the acetone led to a crystalline-to-crystalline phase transition and a green emission. The green-emitting material exhibited a step-wise emission colour-change to yellow, and thence to orange, as a result of ball-milling. The blue-emitting material could be restored by adding acetone to the green-, yellow- or orange-emitting materials. The four emission-states could be switched between at will by acetone-soaking and the use of mechanical stimulation. The blue and green emission materials involved solvated crystalline structures without any defined aurophilic interactions. The yellow-emission powder formed a solvent-free ordered arrangement which involved aurophilic interactions. The amorphous orange-emission material was suggested to involve aurophilic interactions with the shortest gold-gold distance. The transition from green to yellow material required solvent release. The amorphous orange phase was thermodynamically more stable than the yellow phase. The two-step mechanochromism (green to yellow to orange) change during mechanical stimulation began with solvent release from the green material, followed by rearrangement into solvent-free yellow materials via crystalline-to-crystalline phase transition. A crystalline-to-amorphous transition then produced the thermodynamically more stable orange material.

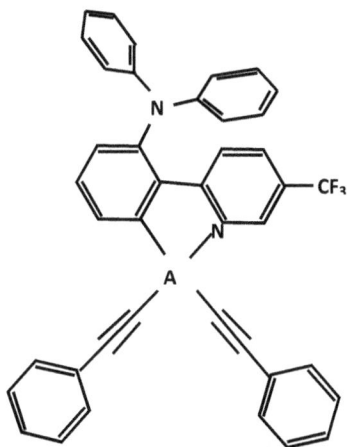

Figure 7. Structure of bis(alkynyl)gold(III) N^C complex C

The complex double salt, $[Au(NHC)_2][Au(C\equiv N)_2]$, has a polymorphic crystalline structure, with various gold-gold distances and ligand-ligand torsion angles[41]. It is a good model system for elucidating structure-spectroscopy correlations and the mechanochromic luminescence of gold complexes. Mechanical grinding of the crystals led to apparent shortening of the gold-gold distance and an increase in the torsion angle. This led in turn to a red-shifted transition.

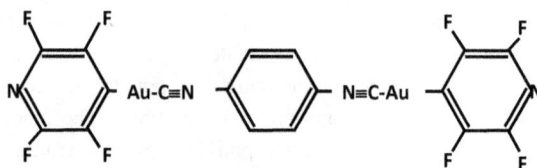

Figure 8. Structure of gold isocyanide complex

Table 2. Colours of gold complexes with various isocyanide ligands

Complex	Condition	λ(nm)
A	as-received	465
A	ground	495
A	fumed	560
B	as-received	466
B	ground	484
B	fumed	466
C	as-received	398
C	ground	489
C	fumed	398

Gold complexes involving various isocyanide ligands have been synthesized[42]. The differing terminal ligands (figure 9) imparted distinctive properties. The alkyl-modified complexes exhibited a tri-colour or bi-colour mechanochromic behavior (table 2) due to a loose packing caused by weak intermolecular C-H···F interactions. The aryl-bridged complexes were not mechano-responsive, due to a tight π-π stacking. The complexes all exhibited differing marked aggregation-induced emissions which were caused by the inherent differences in structure. The third of these compounds[43] forms green-emitting and blue-emitting crystals which exhibit two responses to mechanical stimulation. Upon such stimulation, the emission colour of the green-emitting form does not change but its crystalline structure changes. On the other hand, the blue-emitting form exhibits a clear shift in emission, but it retains its molecular arrangement following grinding.

Gold complexes of the form, ($[Au_6(bis(2\text{-diphenyl}$ phosphinoethyl) phenyl phosphine)$_4$Cl]$(PF_6)_5 \cdot 2(CH_3C_6H_5)$, $[Au_6(bis(2\text{-diphenyl}$ phosphinoethyl) phenyl phosphine)$_4$Cl]$(AsF_6)_5 \cdot 8(CH_3C_6H_5)$ and $[Au_6(bis(2\text{-diphenyl}$ phosphinoethyl) phenyl phosphine)$_4$Cl]$(SbF_6)_5 \cdot 7(CH_3C_6H_5)$, having a box-like shape have been prepared[44]. A chloride ion was located at the center of the box, with two of the six gold ions nearby. The grinding of blue luminescent crystals which contained the, $[Au_6(bis(2\text{-diphenyl}$ phosphinoethyl) phenylphosphine)$_4$Cl]$^{5+}$ cation resulted in their conversion into amorphous solids, exhibiting a green emission, which contained the bridged helicate cation, $[\mu\text{-Cl}\{Au_3(bis(2\text{-diphenyl phosphinoethyl}) phenylphosphine)_2\}_2]^{5+}$.

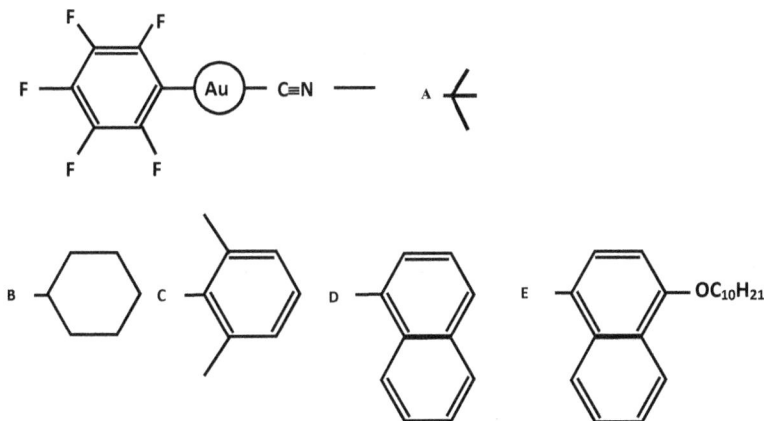

Figure 9. Gold complex with alternative isocyanide ligands

A twisted gold isocyanide complex, based upon tetraphenylethene, exhibited[45] an aggregation-induced phosphorescence which was due to the incorporation of gold and to a conformational rigidity of the aggregate state. The emission colour of the crystalline material changed from 454nm (blue) to 500nm (green) during grinding, due to conformational planarization, enhanced $\pi\cdots\pi$ stacking and the appearance of aurophilic interactions in the post-grinding amorphous state. The original emission colour could be restored by solvent fuming, due to reconstruction of the crystalline lattice. Mechanical stimulation of a 9-anthryl gold isocyanide complex has been found[46] to cause an enormous bathochromic shift of its emission colour from the visible and into the infrared. Before the mechanical treatment, α and β polymorphs had emission-wavelength maxima of 448 and 710nm, respectively. Following grinding, the maximum shifted to 900nm.

Many organic/inorganic and organometallic compounds display mechanochromic luminescence, and most of them undergo a crystalline-to-amorphous phase transition. Examples of crystalline-to-crystalline transformation are rare. A mechanochromic luminescent material which undergoes a crystalline-to-crystalline transformation, mediated by a transient amorphous phase, was produced[47] by introducing soft triethylene-glycol side-chains into a crystalline gold complex which exhibited a mechanochromic luminescence that was based upon a crystalline-to-amorphous phase transition. Mechanochromic luminescence behaviour can be modulated by side-chain substituents which serve as the directing group, as well as substrates which can disrupt the functioning of the directing group[48]. When the gold complex with triethylene-glycol chains was subjected to mechanical or thermal stimulus, irreversible luminescence colour-changes from blue to yellow to green occurred in both cases. The crystallinity of the mechanically generated crystalline phase was inferior to that of the thermally generated crystalline phase. This was attributed to the fact that the mechanically induced conversion-process was completed within seconds, whereas the thermal conversion required several minutes. The crystalline-to-crystalline transformation could be controlled by doping the complex with an inactive soft component. This rendered the transformation reversible, green to blue, upon thermal annealing of the mechanically produced crystalline phase. It was noted that these results suggested the possible development of an imaging process, for covert information storage, that would resist disclosure by ultra-violet illumination.

The luminescent blue complex, $[AuCu_2(NHC)_2(MeCN)_4]^{3+}$, undergoes partial de-solvation upon exposure to air and the emission changes to aquamarine[49]. Grinding produced a yellow emitting material which reverted to blue upon adding acetonitrile. Thermolysis produced a yellow emissive form.

A series of bi-nuclear gold complexes was synthesized[50], and the results of mechanochromic studies suggested that those complexes with methyl-substituted phenyl bridges exhibited mechanochromism. A 100nm red-shift of the fluorescent spectrum was observed after grinding. A complex with a diphenylmethane bridge exhibited mechanochromism with a change in fluorescence from green to blue (25nm red-shift) following grinding. The ground materials reverted to their original states upon treatment with CH_2Cl_2. Bi-isocyano-based bi-nuclear gold complexes were synthesized[51] which differed only with respect to the bridge that linked the two identical arms. All of the complexes exhibited aggregation-induced emission characteristics and mechanochromic behavior. The phosphorescence properties exhibited an off-on switchable green luminescence under grinding. Experimental data indicated that changing multiple intermolecular, C-H$\cdots$F, C$\cdots$F and weak π-π interactions, and aurophilic interactions were critical factors which enabled the aggregation-induced emission and switchable mechanochromic luminescence. Tri-isocyano-based tri-nuclear gold complexes were also synthesized[52] which again exhibited aggregation-induced emission and an off-on green fluorescence under grinding. The same explanation was offered for these phenomena. Seven carbazole-based mononuclear gold complexes having alkyl chains of differing lengths were synthesized[53], and all of them exhibited remarkable aggregation-induced emission characteristics. The addition of alkyl chains to the periphery of the skeleton of organic chromophores is generally done[54] in order to improve their solubility, but it has been shown that the length of the chains could greatly affect solid-state molecular packing modes plus optical and electronic properties; thus helping to tune the solid-state aggregation behavior, opto-electronic properties and the emissions of organic fluorescent materials. The various solid-state light-emitting aggregation-induced emission complexes here also exhibited reversible mechanochromic fluorescence. As above, the mechanism proposed to explain these useful phenomena involved weak multiple intermolecular C-H$\cdots$F and π$\cdots$π interactions, plus the formation or alteration of aurophilic interactions. Further study[55] of these materials confirmed that two of the luminogens exhibited a reversible luminescent colour transformation from yellow-green to green. Two others exhibited a switchable mechanochromic luminescence behavior which involved a fluorescent colour-change from colourless to green. The solid-state fluorescence of a further two luminogens could be switched between weak and strong fluorescence by using mechanical force; but this conversion was irreversible. In a related study[56], one of the luminogens also exhibited a reversible high-contrast mechanochromic and vapochromic luminescence and could emit a persistent room-temperature phosphorescence, with a solid-state emission lifetime of up to 86.84ms. This was then the highest lifetime reported for a gold complex. In fact, it was the first example of an aggregation-induced emission luminogen which exhibited a persistent room-temperature phosphorescence, reversible mechanochromism and vapochromism. A fluorine-based aggregation-induced emission gold complex was also synthesized[57] in which the luminogen exhibited a crystallization-induced emission enhancement effect plus a reversible mechanochromic behavior, with the fluorescence emission changing between green and yellow[58]. When a series of constitutional isomers which contained di-nuclear gold units was synthesized[59], the complexes again exhibited marked aggregation-induced emission but their solid-state mechanochromic properties differed, in that the ortho-isomer luminogen exhibited reversible mechanochromic fluorescence, while the meta-isomer luminogen exhibited a switchable mechanical force-induced luminescence enhancement. No mechanochromism was observed in the case of the para-isomer luminogen. The proposed mechanism here was that morphological phase conversion between crystalline and amorphous states was responsible for the mechanochromism or mechanical force-induced emission enhancement. Five phenylethyne(p-alkoxyphenyl isocyanide) gold complexes, with CH_3, C_2H_5, C_4H_9, C_6H_{13} or $C_{10}H_{21}$ alkyl chains in the para-position of the phenyl isocyanide, were synthesized[60]. As-prepared powder samples of the complexes exhibited differing emission colours. The results indicated the occurrence of weak intermolecular interactions in their crystal structures, and the various alkyl chains had a clear effect upon the stacking arrangements and optical behavior. The luminescence quantum yield of the C_4H_9-chain complex could be as high as 49.3% in the solid state. All of the complexes exhibited luminescence-quenching following grinding. The number of pyridine gold complexes remains quite limited, including those pyridine-containing gold complexes which exhibit aggregation-induced emission or mechanochromism. A relatively new bipyridine-based aggregation-induced phosphorescent emission gold complex[61] exhibits reversible mechanochromism, with the phosphorescent emission changing from weak green to strong yellow. This luminogen also exhibited a marked self-assembly ability. The solid-state phosphorescence and emission lifetimes of mononuclear gold complexes could be tuned[62] by combining fluorophores with triphenylamine, carbazole or tetraphenylethylene. One luminogen exhibited a switchable mechanochromic luminescence, with a colour change from yellow to colourless, another exhibited an on-off solid-state luminescence mechanochromism between yellow-green and colourless. No mechanochromism was observed in the case of tetraphenylethene. The results suggested that the high-contrast mechanochromism of the first two luminogens was due to a crystalline-to-amorphous transition. A highly emissive carbazole-based mononuclear gold complex had a room-temperature phosphorescence lifetime of up to 29.91ms, and exhibited self-reversible mechanochromism[63].

A cholesterol-functionalized gold-isocyanide complex which exhibited mechanochromic luminescence properties was self-organized, into distinct microscopic structures having differing photoluminescence properties, via vapour-diffusion of a poor solvent into dichloromethane solution[64]. The structural-optical relationships of the microstructures were attributed to a mechanically-produced phase transition.

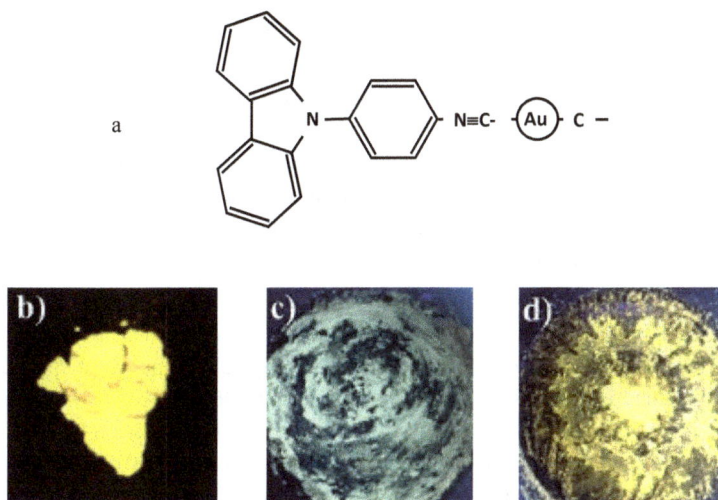

Figure 10. Structure of, and effect of grinding and fuming on, a carbazole-based complex a: structure, b: as-received solid, c: after grinding, d: after fuming

A gold-diphosphine complex was found[65] to exhibit both reversible mechanochromism and mechanochromic luminescence simultaneously. Mechanical grinding could trigger a transformation from a neutral mononuclear structure, which exhibited a white colour and blue photoluminescence, to an ionic di-nuclear structure with intramolecular aurophilic interactions which exhibited a yellow colour and a red emission.

Complexes of the form, $[Au_2(4,6\text{-bis(diphenylphosphino)-phenoxazine})_2]X_2$, where X was NO_3, CF_3COO, CF_3SO_3, $[Au(CN)_2]$ or BF_4, exhibited an anion-switchable and stimulus-responsive luminescence[66]. Depending upon the anion type, the complex could exhibit yellow, orange or red emissions. The reversible mechanochromic luminescence changes were easily visible to the unaided eye. $[Au_2(4,6\text{-bis(diphenylphosphino)-phenoxazine})_2]^{2+}$ cations having short intramolecular gold-gold interactions were involved, as donors, in infinite NH...X hydrogen-bonded chain formation, with X = O and N, plus CF_3COO^- and aurophilically-linked $[Au(CN)_2]^-$ counter-ions. Their room-temperature red and orange emissions became yellow upon cooling to 77K. They exhibited a reversible mechanochromic luminescence, and changed in emission colour from red to dark and from orange to red. Some compounds also exhibited a reversible mechanochromic luminescence, with the emission colours alternating between orange or red and dark, as well as between yellow or orange and red. Here NH...X interactions played a significant role in the stimulus-responsive luminescence.

Grinding of the complex, $[(C_6F_5Au)_2(\mu\text{-}1,4\text{-}di\text{-}isocyanobenzene)]$, produced a reversible photoluminescent colour-change[67] which could be restored by solvent exposure. Such cycling could be repeated 20 times without producing any colour degradation. It was concluded that a change in molecular arrangement was responsible for the mechanochromic behaviour, with intermolecular aurophilic bonding being a key factor governing the altered emission.

A highly emissive carbazole-based mono-nuclear complex (figure 10a) has been synthesized[68] which exhibited a bright yellow luminescence and a room-temperature phosphorescence lifetime of up to 29.91ms. The original luminescence spectrum of the solid contained emission bands with maxima at 436, 460, 546 and 594nm. Following grinding, the yellow-emission changed to yellow-green, with emission bands at 468, 546 and 594nm. Fuming with dichloromethane solvent vapor for 30s caused the original intense yellow luminescence to reappear (figures 10b, c, d).

Most recently, benzo[b]phosphole alkynylgold(I) complexes, prepared by incorporating photochromic dithienylethene-containing benzo[b]phosphole into the triphenylamine-containing arylethynyl ligand, have been shown[69] to exhibit mechanochromic properties. A good fatigue resistance, repeatable cycles without any discernible loss of reactivity and multi-color states controlled by mechanochromism, are observed. The above strategy for the molecular design of these materials is expected to expand the future exploitation of photo-responsive gold(I) complexes.

Copper Complexes

Establishing the mechanism responsible for mechanochromism is problematic, due to the difficulty of characterizing the ground phase. A study[70] of the real crystalline polymorphs of a mechanochromic luminescent copper iodide cluster permitted the clear identification of the mechanism which was involved. That is, local disruption of the crystal packing produced changes in the cluster geometry and modification of the cuprophilic interactions; thus modifying the emissive states.

A comparison was made[71] of two copper iodide compounds which differed with regard to a subtle modification of the ligands. The two clusters had very similar crystalline structures but very different optical properties; only one exhibited luminescence mechanochromism. The two types of cluster had very different internal stresses, due to slight differences in the crystal packing, which were in turn controlled by the nature of the ligands. The release of those constraints during mechanical treatment led to modification of the intramolecular interactions and thence to the mechanochromism.

Di-nuclear copper complexes with hexadentate macrocyclic N-heterocyclic carbene ligands, $[Cu_2(L^1)(CH_3CN)][PF_6]_2$ and $[Cu_2(L^2)(CH_3CN)]_2[Cu_2(L^2)(CH_3CN)_2][PF_6]_6$, have been synthesized[72] by reacting $[H_4L^1][PF_6]_4$ or $[H_4L^2][PF_6]_4$ with excess Cu_2O in acetonitrile. Crystallization of heat-treated samples of $[Cu_2(L^1)(CH_3CN)][PF_6]_2$ and $[Cu_2(L^2)(CH_3CN)]_2[Cu_2(L^2)(CH_3CN)_2][PF_6]_6$ from acetone/methanol/ether or CH_3NO_2/ether, resulting in $[Cu_2(L^1)][PF_6]_2$ and $[Cu_2(L^2)][PF_6]_2$. All of the complexes were emissive, with luminescence maxima at 464, 472, 540 and 488nm, respectively, in the solid state. The red-shift of the emission maximum of $[Cu_2(L^1)][PF_6]_2$ relative to the other three complexes was attributed to cuprophilic interactions in its excited state. Following grinding, an appreciable emission colour-change was observed, with a red-shift of 98nm for $[Cu_2(L^1)(CH_3CN)][PF_6]_2$, 82nm for $[Cu_2(L^2)(CH_3CN)]_2[Cu_2(L^2)(CH_3CN)_2][PF_6]_6$, 20nm for $[Cu_2(L^1)][PF_6]_2$ and 64nm for $[Cu_2(L^2)][PF_6]_2$. These mechanochromic transformations were crystalline-to-amorphous conversions and could be reversed by adding drops of organic solvent or by recrystallization. There were possible correlations between the luminescent properties and changes in the Cu⋯Cu separations.

Two luminescent CuI coordination polymers, $[Cu_2I_2(triphenylphosphine)_2(1,3,5-tris(3-pyridyl)benzene)]_n$ and $[Cu_2I_2(triphenylphosphine)_2(1,3,5-tris(4-pyridyl)benzene)]_n$ have been synthesised[73], which contained one-dimensional coordination chains with rhombic $\{Cu_2I_2(triphenylphosphine)_2\}$ cores and 1,3,5-tris(3-pyridyl)benzene or 1,3,5-tris(4-pyridyl)benzene bridging ligands. Both the Cu-1,3,5-tris(3-pyridyl)benzene and Cu-1,3,5-tris(4-pyridyl)benzene exhibited a blue-to-yellow thermally activated delayed fluorescence which arose from a mixing of the metal-to-ligand and halide-to-ligand charge-transfer excited states; giving and emission quantum yields of 0.29 and 0.27, respectively, at 298K. Mechanochromic luminescence was observed for both complexes. The emission lifetimes showed that the cause of the emission switched from thermally-activated delayed fluorescence to phosphorescence, and arose from a triplet cluster-centered emissive state that was generated by grinding-induced amorphization.

Di-nuclear copper complexes were prepared[74] by reacting PNP ligands, 2,6-$Me_2C_6H_3N(triphenylphosphine)_2$, with CuCl and CuBr. By varying the position of the methyl substitution from ortho to meta, tri-nuclear 3,5-$Me_2C_6H_3N(triphenylphosphine)_2$ copper complexes were obtained. Methyl substitution at the para position produced tetra-

Materials Research Forum LLC
https://doi.org/10.21741/9781644900277

nuclear 4-MeC$_6$H$_4$N(triphenylphosphine)$_2$ copper complexes. Single-crystal X-ray studies showed that the Cu$\cdots$Cu separations in the various complexes ranged from 2.650 to 2.935Å. The copper complexes, 3,5-Me$_2$C$_6$H$_3$N(triphenylphosphine)$_2$ and 4-MeC$_6$H$_4$N(triphenylphosphine)$_2$, exhibited mechanochromic as well as thermochromic luminescences which were attributed to a decrease in their Cu$\cdots$Cu separations.

Flexible films of composites based upon organometallic coordination compound, copper-phenylacetylide, nanobelts and single-wall carbon nanotubes exhibit[75] an unusual mechanochromic luminescence change from bright green to dark red following grinding of the crystalline copper-phenylacetylide nanobelts. The highly flexible composites also have room-temperature thermoelectric power-factors of up to 200.2µW/m^{2}K.

Two copper-halide based coordination polymers, [Cu$_3$Cl$_2$(tris(4-(1H-imidazol-1-yl)phenyl)amine)$_2$]Cl and [Cu$_3$I$_3$(CuI)$_2$(tris(4-(1H-imidazol-1-yl)phenyl)amine)$_3$], were prepared[76] using solvothermal reactions. The former compound had a chain structure, with unique Cu$\cdots$Cu interactions, C-H$\cdots$Cl halogen-hydrogen bonds and $\pi\cdots\pi$ stacking. It exhibited unusual mechanochromic properties.

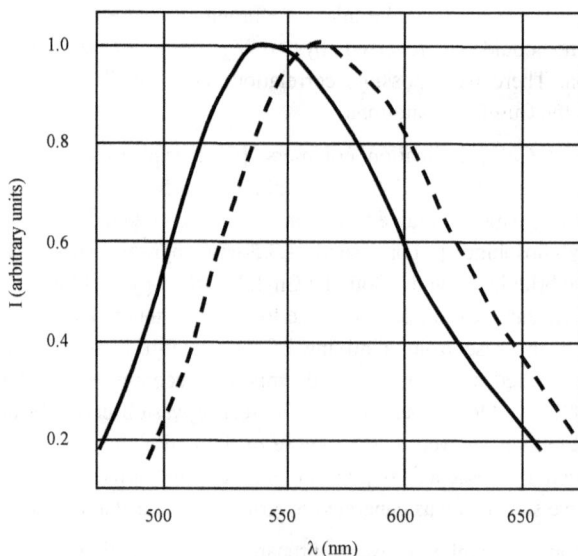

Figure 11. Normalized solid-state emission spectra for unground (solid line) and ground (dashed line) [CuBH(benzimidazolate)$_3$]$_n$ samples

The copper coordination polymer, $[Cu_2I_2L^2]_n$, exhibits[77] luminescence mechanochromism, with a colour-change from greenish-blue to yellow occurring upon applying pressure. Changes in the bond lengths were the main cause of the mechanochromism and the crystallinity was preserved, with small lattice distortion, in spite of the very high pressure. It was thus a non phase-transition process. Upon adding drops of acetonitrile to ground and compressed samples, the original greenish-blue emission and crystalline state were restored, thus making the colour-change fully reversible.

A copper iodide compound, catena(bis(μ2-iodo)-6-methylquinoline-copper, was synthesised[78] which comprised isolated polymeric staircase chains of copper-iodide that were coordinated to organic ligands via Cu-N bonds. High-pressure (up to 6.45GPa) X-ray diffraction revealed that the material was soft, with a bulk modulus of 10.2GPa. The unit-cell compression was very anisotropic, with [302] being the stiffest direction due to a combination of stiff CuI ladders and the mutual shear of planar quinolone ligands. High pressures reduced the Cu$\cdots$Cu separations and this was reflected, in the luminescence spectra, by the appearance, at above 3.5GPa, of four sub-bands at 515, 600, 647 and 712nm. Red-shifts were attributed to interactions, between the copper ions, due to shortening of the Cu$\cdots$Cu separations which were reduced, by pressure, to below twice the van der Waals limit of 2.8Å.

Low-dimensional copper boron imidazolate[79,80] frameworks, $[CuBH(benzimidazolate)_3]_n$ and $[CuBH(imidazolate)_3]_n$, were synthesized[81], both of which exhibited mechanochromism. The room-temperature solid-state photoluminescent properties of the former material were explored by irradiation with ultra-violet light at 360nm. The maximum emission wavelength of the material was 548nm, and this was attributed to [Cu $\rightarrow \pi^*BH(benzimidazolate)_3^-$] metal-to-ligand charge transfer. Following grinding, the emission spectrum exhibited a clear red-shift, of about 20nm, to 568nm (figure 11); a behaviour which is almost universal for mechanochromic materials. There were only very weak and ambiguous crystallographic reflections, indicating that the lattice was significantly disrupted and that a crystalline-to-amorphous conversion was caused by strong grinding. Because the material has a double-chain structure, disordered sliding between chains, alterations of Cu$\cdots$Cu interactions, C–H$\cdots\pi$ interactions and even $\pi\cdots\pi$ interactions between the chains, were linked to the luminescence change.

A general study[82] of the mechanochromism of copper iodide clusters under hydrostatic pressures which were sufficient to control the crystal packing via modification of the intermolecular interactions, established a direct correlation between molecular structure and luminescence. In particular, cuprophilic interactions were shown to be responsible for the stimulus-responsive luminescence of such multi-nuclear coordination compounds.

A new mechanism, mechanically-induced reversible formation of a cation-anion exciplex based upon Cu-F interactions, has been proposed[83] which leads to highly efficient mechanochromic phosphorescence and to unusually large emission shifts from ultra-violet blue to yellow in CuI complexes. The low-energy luminescence is both thermo-responsive and vapo-responsive, thus permitting the generation of white light and recovery of the original blue emission.

A significant enhancement of luminescence intensity has been demonstrated[84] in 1-dodecanethiol-capped copper nanoclusters, due to self-assembly. Upon forming compact ordered assemblies, originally non-luminescent copper nanoclusters exhibit strong emissions. Self-assembly also permits control of the polymorphism of copper nanocluster assemblies and thus of the emissions. A relationship was found between the compactness of the assembly and its emission. A high compactness enhances the cuprophilic Cu···Cu inter- and intra- interactions of nanoclusters and suppresses intramolecular vibration and rotation of the capping ligand of dodecanethiol. The emission energy depends upon the Cu···Cu separation and the increased compactness increases the average Cu···Cu separation by introducing additional inter-nanocluster cuprophilic interactions, leading to a blue-shift of the nanocluster emission. The nanocluster assemblies therefore exhibit a distinct mechanochromic luminescence.

Reaction between CuI and methyl or methyl-2-amino-isonicotinate leads[85] to the formation of two coordination polymers which are based upon double zig-zag Cu_2I_2 chains. The presence of a NH_2 group in the isonicotinate ligand leads to supramolecular interactions which affect the Cu-Cu distances, the symmetry of the Cu_2I_2 chains and the physical properties. Both of the above coordination polymers are semiconductors and exhibit a reversible mechanoluminescence. A preparation method based upon rapid precipitation produces nanofibers of the above coordination polymers. The dimensions of the nanofibers permit the preparation of mechanochromic film composites, with polyvinylidene difluoride, which are tens of nm in thickness and some cm in length.

Studies of luminescent copper complexes[86] have revealed that some of them exhibit chromic luminescence under various external stimuli. Tetra-nuclear copper clusters having a cubane-type Cu_4I_4 core are the most notable group, and their response to stimuli depends greatly upon coordinating organic ligands. Coordination-polymerization of the copper-cluster cores using an organic link imparts chromic behaviors which depend strongly upon the flexibility of the framework. The mechanochemical synthesis of highly luminescent copper complexes is closely related to mechanochromic luminescence.

The isomer, [Cu$_4$I$_4$(triphenylphosphine)$_4$], of the most common cubane form of copper iodide cluster has a chair geometry. A molecular cluster having the formula, [Cu$_4$I$_4$(triphenylphosphine)$_4$]$_2$CHCl$_3$, exhibits a high-contrast emission response to grinding, and the optical properties of the ground phase closely resemble those of the cubane isomer. Following grinding, the solid-state mechanochromic luminescence of Cu$_4$I$_4$[triphenylphosphine(CH$_2$CH=CH$_2$)]$_4$ was markedly affected[87]. This reversible effect was attributed to distortions of the crystal packing, thus leading to modification of the intermolecular interactions and of the Cu$_4$I$_4$ cluster geometry. Modification of the Cu-Cu interactions seemed to change greatly the emissive properties of the cluster. Comparisons were made[88] with two related compounds. One was a pseudopolymorph of the form, [Cu$_4$I$_4$(triphenylphosphine)$_4$]CH$_2$Cl$_2$, which also exhibited luminescent mechanochromism. The other was a chair compound, but one containing a slightly different phosphine ligand, [Cu$_4$I$_4$(diphenylphosphineC$_6$H$_4$CO$_2$H)$_4$]. This one did not exhibit mechanochromism but, in general, the materials exhibited a mechanochromism which was based upon solid-state chair $\rightarrow$ cubane conversion. Clusters having a Cu$_4$I$_4$ cubane core had previously been modified[89] by adding phosphine ligands bearing proto-mesogenic gallate-based derivatives in the form of long alkyl chains (C$_8$, C$_{12}$, C$_{16}$) or cyano-biphenyl fragments. The clusters which were modified using only long alkyl chains had an amorphous or crystalline form. Clusters which comprised cyano-biphenyl fragments had the liquid-crystal form of a smectic-A mesophase between room temperature and 100C. The cyano-biphenyl derivative exhibited an unusual luminescence thermochromism which was attributed to resonance-energy transfer between the emissive Cu$_4$I$_4$ inorganic and cyano-biphenyl components. The emission properties of this cluster were sensitive to variations in the local order of the molecular arrangement. The liquid-crystalline properties which were imparted to the inorganic core permitted easy deformation of its environment and led to a mechanochromism which was related to a modulation of the intramolecular interactions. Mechanical constraints imposed on the self-assembled molecular structure provoked changes at the molecular level by modifying the Cu$_4$I$_4$ inorganic core geometry, and especially the strength of cuprophilic interactions.

Two different crystalline polymorphs were found[90] to exhibit distinctly different luminescences, with one being green and the other one yellow. Following grinding, only one of the polymorphs suffered appreciable modification of its emission; from green to yellow. The properties of the resultant partially-amorphous crushed compound were close to those of the other yellow polymorph. A correlation was found between the Cu-Cu bond distances and the luminescence. Probing of the local structure of Cu$_4$I$_4$P$_4$ cluster cores, using ^{31}P and ^{65}Cu, revealed that grinding modified the environments of the phosphorus and copper atoms. The mechanochromism was therefore attributed to disruption of the crystal packing and to a shortening of the Cu-Cu bond distances in the Cu_4I_4 cluster core; thus modifying the emissions. This confirmed the influence of cuprophilic interactions in the mechanochromism of copper iodide clusters.

Two isostructural metal-organic framework materials, $\{[MeSi(3\text{-pyridyl})_3]_6(Cu_6I_6)\}_n$ and $\{[MeSi(3\text{-quinolyl})_3]_6(Cu_6I_6)\}_n$, were synthesized[91] from tridentate arylsilane ligands of the forms, $MeSi(3\text{-pyridyl})_3$ and $MeSi(3\text{-quinolyl})_3$, respectively. The 3-pyridyl version exhibited the thermochromism which was to be expected of the usual $Cu_4I_4(3\text{-pyridyl})_4$ clusters. The other version exhibited emissions which were due to the Cu_6I_6 cluster cores at 298 and 77K. An unusual reversal of the mechanochromic luminescent behavior was observed at 298K. A markedly blue-shifted (orange to yellowish-orange) and high-energy emissions were found for $\{[MeSi(3\text{-pyridyl})_3]_6(Cu_6I_6)\}_n$ whereas a red-shifted (green to orange) low-energy emission was found for $\{[MeSi(3\text{-quinolyl})_3]_6(Cu_6I_6)\}_n$ due to grinding. This was attributed mainly to variations in their cuprophilic interactions, in that $\{[MeSi(3\text{-pyridyl})_3]_6(Cu_6I_6)\}_n$ had shorter (2.745Å) Cu$\cdots$Cu distances than did $\{[MeSi(3\text{-quinolyl})_3]_6(Cu_6I_6)\}_n$ (3.148Å). Ground samples of the latter material thus exhibited a marked red-shift in luminescence due to the reduction of its Cu$\cdots$Cu separations; bringing the value closer (2.80Å) to the der Waals radii between two copper atoms. The blue-shifted emission of $\{[MeSi(3\text{-pyridyl})_3]_6(Cu_6I_6)\}_n$ was attributed to a rise, in the lowest unoccupied molecular orbital energy levels, due to changes in the secondary packing forces.

Although a polymer is not, strictly speaking, a complex it is instructive to include the following case here. Three two-dimensional luminescent isomeric porous coordination polymers were synthesized[92] which were based upon Cu(5-(4-pyridyl)tetrazole). When crystalline samples were gently ground, the colours of all of the materials in ambient light remained essentially unchanged. Upon irradiation with 365nm (ultra-violet) light, samples which incorporated ethanol, toluene and 5-(4-pyridyl)tetrazole appeared as green, yellow and orange, respectively. When the samples were strongly ground, the luminescence were all relatively weak orange ones; consistent with the corresponding emission spectra. The lifetimes of the emissions were very similar for all of the materials and grinding conditions. X-ray powder diffraction patterns of the as-received crystalline samples indicated that their structures did not change during gentle grinding, although prominent peaks disappeared when the samples were strongly ground, indicating conversion to an amorphous state. The latter was attributed to disordered sliding between the Cu(5-(4-pyridyl)tetrazole) layers of the structures. The sliding was assumed to be accompanied by alterations in the Cu$\cdots$Cu and π-π interactions; thus affecting the photoluminescence properties. X-ray diffraction patterns of the ground samples did not contain prominent peaks after the samples had been treated with toluene or ethanol for days, and this was ascribed to insolubility of the coordination polymers in the organic solvents; so that recrystallization barely occurred. On the other hand, ground samples of the material which incorporated ethanol immediately exhibited a bright yellow luminescence under ultra-violet light when a single drop of ethanol was added. No analogous behaviour was exhibited by the other materials. Solid-state photoluminescence measurements of ground samples of the material which incorporated ethanol, when treated with ethanol, has a broad emission band which was centered on 562nm when excited at 296nm. Thermogravimetry revealed that guest molecules of ethanol were lost during grinding. Such losses were expected to leave voids for subsequent occupation by ethanol molecules, thus inducing the glide of Cu(5-(4-pyridyl)tetrazole) layers and consequent luminescence changes. When the sample was completely dry, the luminescence changed to dark-orange, and this cycle could be repeated at least 3 times. Thermogravimetry of the other materials, before and after grinding, did not detect any appreciable changes; suggesting that toluene guest-molecules were only partially removed by grinding. This was in turn attributed to the higher boiling point of toluene as compared to that of ethanol. Each of the ground sample materials was immersed in ethanol, chloroform and toluene. Recovery in chloroform occurred after one month. The luminescence properties recovered completely for samples immersed in chloroform. These results were explained in terms of the inherent self-restorative capability of coordination polymers in solvents.

Platinum Complexes

The H-shaped pentiptycene structure can offer mechanochromic properties, as illustrated by platinum complexes in which bromine or pentiptycene acetylene is a substituent on the terdentate dipyridylbenzene N^C^N ligand, and chlorine or pentiptycene acetylene is an ancillary ligand. Intermolecular interactions between the planar NCN-Platinum cores are governed by π-π and d$-\pi$ interactions, with negligible Pt$\cdots$Pt bonding; corresponding to ligand-centered excimer, rather than metal-metal-to-ligand charge-transfer emission, in the solid state. An interplay between relative excimer-to-monomer emission intensity under an imposed force accounts for a luminescence mechanochromism of pentiptycene-incorporated platinum complexes, whereas a pentiptycene-free complex exhibits essentially no emission colour response. A two-stage emission colour-change from red to orange, and then to yellow, occurred[93] as a result of grinding and subsequent benzene-vapor fuming.

Platinum complexes which featured[94] pyridine bis-N-heterocyclic-imidazol-2-ylidene/-mesoionic-triazol-5-ylidene donors as pincer ligands and chloro, acetonitrile or cyano groups as auxiliary ligands could be prepared as highly strained organometallic phosphors. Four of the complexes had a distorted square planar structure in which the pincer ligand and its mesityl wing-tips were arranged in a twisted conformation relative to each other. The emission response under mechanical shear of a platinum complex with pyridine bis-imidazol-2-ylidene and a weakly-donating acetonitrile auxiliary ligand involved a mechanochromic colour change from sky-blue to green or yellow-green.

Complexes of the form, $[Pt(C^{\wedge}N)(2,2',6,6'\text{-tetramethylheptane-3,5-dionate})]$[95], are such that modifying the phenyl ring of the $C^{\wedge}N$ ligand so as to incorporate an $-SF_5$ group has marked effects on the properties. In a meta position with respect to the $Pt\text{-}CC^{\wedge}N$ bond, the substituent has a largely stabilising effect upon the lowest triplet excited state, which red-shifts the emission of the complex with regard to $[Pt(2\text{-phenylpyridinato})(2,2',6,6'\text{-}$ tetramethylheptane-3,5-dionate)]. When the $-SF_5$ group is located at a para position relative to the $Pt\text{-}CC^{\wedge}N$ bond, it does not directly affect the triplet state, and the group stabilises the metal-based orbitals thus resulting in a blue-shift of the emission. All of the complexes are mechanochromic in the solid state and can exhibit an excimeric emission which originates from intermolecular $\pi\text{-}\pi*$ interactions. The relative emission intensities of the monomeric and dimeric excited states are related to steric congestion of the metal centre.

A pincer platinum complex with amide groups has been investigated[96] with regard to changes in the luminescence colour in the solid state. The complex exhibited mechanochromic behavior and the emissions changed from green to orange during grinding.

Enantiomeric chiral alkynylplatinum bipyridine complexes have been prepared[97] which exhibit luminescent mechanochromism. Mechanical treatment could cause a colour-change from yellow to orange and a luminescence variance from orange to red. During mechanochromism due to grinding, X-ray diffraction patterns became broad and weak and distinct non-linear optical activity of the crystallites vanished, indicating crystalline-to-amorphous and ordered-to-disordered transformations. Enantiomeric chiral cyclometalated complexes with 4,5-pinene-6'-phenyl-2,2'-bipyridine have also been studied[98,99] which underwent a crystalline-to-amorphous and ordered-to-disordered transformation during grinding. In these grinding-produced aggregates, the luminescence was red-shifted by 160nm, as compared with that of crystallites. The electronic circular dichroism signals and non-linear optical responses became negligible. This amorphous phase could be easily restored to the original crystalline state by treatment with a few drops of dichloromethane.

A platinum complex which has an oligo-ethyleneoxide pendant is able to self-assemble into ultra-long ribbons which exhibit mechanochromism under nanoscale mechanical stimulation produced by an atomic force microscopy probe[100]. This reveals that nanoscale mechanochromism arises from static pressure (piezochromism) and from shear-based mechanical stimulus (tribochromism). Confocal imaging shows that mechanochromism occurs only within a short distance from the stimulation.

The crystal structure of the di-nuclear platinum complex, with bis(salen) and 2-methylenepropane-1,3-diyl as the ligand and linker, respectively, is such that the two salen units are almost independent and the linker has but a minimal effect. The solid-state ultra-violet visible spectra of the complexes have a weak absorption band, at 520nm, which is absent from solution spectra. The weak band is attributed to the stacked molecular structure of the complex. Under mechanical treatment, microcrystals and deformed thin films of one form of the complex exhibit photo-emissions of various colours, while another form of the complex exhibits no appreciable emission[101]. Changes in the emission colour are attributed to the semi-flexible linker, which produces a fragility of the molecular packing of the crystals under shear stresses.

Square planar complexes, based upon modified 2-phenylpyridine derivatives as the main ligand, and picolinic acid as the auxiliary ligand have been synthesized[102]. Only those complexes in which a single long flexible chain was attached to each end of the 2-phenylpyridine exhibited monotropic smectic liquid-crystalline behavior. Mechanochromism was observed upon grinding the solid complexes. It was inferred that the liquid-crystalline and mechanochromic properties of the complexes were related to the number of flexible alkoxy chains in the cyclometalated ligand.

Reversible mechanochromic luminescence is observed[103] in cationic platinum terpyridyl complexes. In the solid state the complexes, [Pt(4'-(p-nicotinamide-N-methylphenyl) - 2,2':6',2''-terpyridine)Cl]X$_2$, exhibit colours under ambient light which range from red to orange to yellow. All of the complexes are brightly luminescent at room temperature and at 77K. During grinding, the yellow complexes (X = ClO$_4$ or BF$_4$) turn orange and exhibit a bright red luminescence. This can be restored to yellow by adding a few drops of acetonitrile. Crystallographic studies of the yellow and red forms suggest that, in at least one case, the mechanochromism is due to a change in the intermolecular Pt···Pt distances during grinding.

The square-planar di-imine complex, Pt(4-bromo-2,2'-bipyridine)(C≡CC$_6$H$_5$)$_2$, was studied[104] showing that the solid state exhibited reversible grinding-provoked colour and luminescence changes. When heated or ground, the original bright yellow-green emission changed by between 88 and 165nm while the crystalline state became amorphous. When the amorphous material absorbed organic vapour, it could revert to the crystalline state while the red luminescence returned to yellow-green.

The square-planar di-imine complex, Pt(4-[(2-trimethylsilyl) ethynyl] -2,2'- bipyridine) $(C{\equiv}CC_6H_5)_2$, was found[105] to exhibit reversible mechanochromism. When the crystalline material was heated or ground, the bright yellow emission changed by a red-shift of up to 178nm. When the heated or ground samples were exposed to organic vapours, they reverted to the crystalline state and the luminescence changed from red to yellow. The change was attributed to charge-transfer.

Iridium Complexes

Complexes having various lengths of alkyl chain were synthesized[106], revealing that all of them exhibited an easily visible and reversible mechanochromic behavior which depended upon the chain length: the shorter the alkyl chain, the more marked was the mechanochromism and the higher was the recrystallization temperature. This was attributed to interconversion between the crystalline and amorphous states under external disturbance. Changing the alkyl chain had a negligible effect upon the emission colour. Theoretical calculations suggested that an intrinsic intramolecular π-π stacking tended to restrict opening of the structures in the metal-centered excited state.

Ionic complexes[107] which exhibited mechanochromic, vapochromic and electrochromic phosphorescence simultaneously shared the same phosphorescent iridium cation with a N-H group in the N^N ligand and contained various anions; such as hexafluorophosphate, tetrafluoroborate, iodide, bromide and chloride. The anionic counter-ions produced a variation in the emission colours of the complexes, from yellow to green, due to the formation of hydrogen bonds with the N-H proton. The N-H group was sensitive to grinding, vapours and electric fields, thus leading to the above-noted phosphorescences.

The photophysical properties of series of cationic iridium phosphors that contained the same main ligands, but easily variable donor-acceptor ancillary ligands, have been explored[108]. All of the materials exhibited intense emissions in the solid state, due to intrinsic aggregation-induced emission. By changing the donor components on ancillary ligands, controllable and clearly visible mechanochromic luminescence was obtained. The slight modification of donor groups on ancillary ligands led to clear changes in the dipole moments, with larger dipole moments leading to more efficient mechanochromism. The emission colour of aggregation-induced emission iridium complexes could be varied[109] by adjusting the donor/acceptor strength on ancillary ligands. The complexes all involved 1-(2,4-difluorophenyl)-1H-pyrazole as a cyclometalated ligand, but had differing pyridine-1,2,4-triazolyl components as ancillary ligands which were modified using carbazole end-capped alkyl groups in which pyridine-1,2,4-triazolyl and carbazole components acted as donor and acceptor units, respectively.

There was an intrinsic relationship between the structure and the emission behavior, in that enhancing the influence of donors and/or acceptors led to red-shifted emissions. One of the materials exhibited an appreciable reversible mechanochromic luminescence due to efficient aggregation-induced emission.

Cationic cyclometalated complexes of the form $[(C\char`^C)IrX(triphenylphosphine)_2(4,4'-dimethoxy-2,2'-bipyridine)]A$, where X was a halogen and A was a counter-ion, have been synthesised[110]. Changing the counter-anion or halogen ligand altered the photophysical properties. The complex in which X and A were both chlorine did not exhibit any mechanochromism, whereas other choices did; with bright solid-state phosphorescence and a reversible room-temperature mechanochromic luminescence.

Other Metal Complexes

A silver complex with o-bis(diphenylphosphino)benzene was found[111] to exhibit reversible mechanochromic luminescence. The colourless crystals which resulted from recrystallization from $CHCl_3$ ether were blue under ultra-violet irradiation and, when ground, the resultant white powder emitted a green luminescence. Upon treatment with drops of $CHCl_3$ hexane, or heated (600s, 200C) the previous state was restored. There was a reversible conversion from crystalline to amorphous during the grinding and heating. It was supposed that grinding disrupted the intermolecular interactions between the phenylene rings of adjacent molecules, resulting in the green emission. A hexa-nuclear complex with $\{[Ag_3(iPrNHC(S)NP(S)(OiPr)_2-S,S')_3]_2\}$ exhibited[112] a reversible conversion, between a yellow-emitting form and a blue-emitting form, during grinding and recrystallization.

It is perhaps stretching a point to include bismuth under this heading, but it is convenient given that the chemistry is essentially the same as that for metal complexes. The α and γ polymorphs of $BiBr_3(N-oxide-2,2'-bipyridine)_2$ are aggregation-induced emission phosphors which exhibit[113] mechanochromism. Grinding produces a crystalline-to-amorphous transition and a change in luminescence from green (525nm for α and 503nm for γ) to red (611nm for α and 614nm for γ). These changes are reversed by heating, by fuming or by recrystallization. Similar studies[114] of two coordination polymers, (tetrabutylammonium)$[BiBr_4(N-oxide-4,4'-bipyridine)]$ and $[BiBr_3(N-oxide-4,4'-bipyridine)_2]$, revealed that the emissions were due to aggregation-induced phosphorescence. Upon grinding, both materials became amorphous, and the luminescence changed from yellow to orange in the case of the former material and from orange to red in the case of the latter. Annealing, or exposure to water vapor, recovered the original luminescence.

Two-dimensional layered metal–organic micro/nanosheets of [Cd(anthracene-9-carboxylate)$_2$(benzimidazole)$_2$] and its solvate CH_3CN-containing form exhibit reversible mechanochromic delayed fluorescence. Upon applying a force, the luminescent center of the solvate form could be converted from an anthracene-9-carboxylate/benzimidazole exciplex to an anthracene-9-carboxylate/anthracene-9-carboxylate excimer, resulting in alternations of the delayed fluorescence[115]. The luminescence change could be recovered by CH_3CN fumigation. An alkaline-stable anionic boron imidazolate framework, CdB(imidazolate)$_4$(benzene-1,2,4,5-tetracarboxylate)0.5H_2O, has been synthesized[116] which has a three-dimensional structure and exhibits mechanochromism with a maximum emission blue-shift of about 23nm in the blue-emitting range. A mechanoresponsive luminescent Cd coordination polymer has been synthesized[117] which exhibits a sensitivity to mechanical stress such that the luminescence colour changes gradually from weak yellowish-green to bright cyan during grinding.

The three-dimensional hydrogen-bonded complex, {[Co(3-(4-pyridyl)benzoate)$_2$(H_2O)$_4$}$_n$, exhibits[118] a mechanochromic behaviour between 30 and 120C which is attributed to disruption of the supramolecular arrangement or to metal-to-ligand charge-transfer. At above 150C, it is attributed to a loss of coordinated water molecules from the cobalt ions, which then alters the visible electronic transitions. Following the grinding of crystals, a purple colour is observed in ambient light and this mechanochromism is attributed to changes in the crystal field around the cobalt centre. These effects are fully reversible.

The metal-organic compound, {[Eu(Bpebc)$_{1.5}$(SO_4)(H_2O)$_2$](OH)•16.5H_2O}$_n$, where H_2BpebcCl$_2$ = 1,10-bis-(4-carboxybenzyl)-trans-1,2-bis(4-pyridinium)ethylene dichloride), has a Borromean-ring type structure and exhibits[119] mechanochromic luminescence, with on-off switching. The excitation wavelength-dependent emission behavior can be used to vary the emission colour from violet to orange. This is attributed to a pressure-induced shift of the lattice faces, which weakens intermolecular interactions.

The complexes, [Fe(2,2'-dimethyl-4,4'-bithiazole)$_3$][FeCl$_4$]$_2$ and [Fe(2,2'-dimethyl-4,4'-bithiazole)$_3$][FeBr$_4$]$_2$, comprised[120] high spin iron centers in the octahedral cation part of the complex, with an Fe-N distance of 2.220Å, in spite of the low-spin octahedral iron complexes with unsubstituted bithiazole ligands. A temperature reduction to 90K resulted in a decrease in the Fe-N bond length to 2.206Å. The chlorine-containing complex exhibited a reversible mechanochromic change from red to yellow in going from crystalline to powder form.

In the PbX$_2$/N-oxide-4,4'-bipyridine system[121], four coordination polymers of the form, [PbX$_2$(N-oxide-4,4'-bipyridine)], were found. When X was chlorine or bromine, they were polymorphs having acentric structures. When X was iodine, the compound had a similar but centrosymmetric structure. When X was NO_3, it consisted of 2 interpenetrating three-dimensional networks. All of the materials had phosphorescence properties which included a broad emission band at about 600nm, with lifetimes that were greater than tens of μs. They also exhibited mechanochromic luminescence, in that grinding led to almost complete extinction of the luminescence, together with a crystalline-to-amorphous transition. This could be reversed by heating, exposure to water or acetone vapour or by recrystallization in a few drops of acetone. This however led to a loss of emission intensity.

A bis(salicylaldiminato)Zn Schiff base complex, derived from benzo[c][1,2,5]thiadiazole-5,6-diamine, has been synthesized[122] which exhibits a force-induced luminescence change. During grinding, as-prepared solid which was crystallized from ethanol/dichloromethane solution exhibited an emission-colour change from yellow (545nm) to red (645nm). This change could be removed by fuming, and the yellow-red change repeated. Annealing of as-prepared solid produced a more ordered orange (575nm) phase. Mechanical force could produce a morphological transformation from crystalline to amorphous phase, accompanied by a change in the material's molecular packing. The yellow emitting as-prepared solid had a columnar square molecular arrangement. The complex was able to form organic luminescent gels which were made up of one-dimensional molecular nanofibrils. A xerogel form of the material also exhibited mechanochromism.

Tetraphenyls

The luminescence-reversal behaviour of tetraphenylethene analogues is of great interest. Mechanical treatment can not only destroy the crystallinity, but can also have a marked effect upon the amorphous-to-crystalline transition of amorphous materials. Grinding-enabled recrystallization annealing is attributed to the presence of twisted and rotatable intramolecular conformations which can be altered by mechanical stimuli. Mechanochromic materials which exhibit aggregation-induced enhanced emission are of particular interest. Intermolecular interactions affect photophysical processes, but their role in the luminescence of aggregation-induced enhanced emission is difficult to identify because pressurized samples are often amorphous. Unusual pressure-dependent changes in tetraphenylethene have been detected and linked to the flexible role played by aromatic C-H⋯π and C-H⋯C components. In order to study the effects of donor and acceptor substitutions on tetraphenylethene, three typical cases were examined[123]. One film sample exhibited a green fluorescence (494nm), aggregation-induced emission and a reversible mechanochromism from blue (472nm) to green (505nm), in grinding and fuming cycles.

The replacement of two phenyl groups by two cyano groups produced a film which exhibited orange fluorescence (575nm) and aggregation-induced emission. The mechanochromic behavior, from yellow (541nm) to orange (563nm), of another derivative was reversible under grinding and fuming or grinding and annealing cycles.

The combination of achiral tetraphenylethene, an aggregation-induced emission enhancement luminophore with four long alkyl chains, permits - via amide linkage - the control of supramolecular chirality so as to construct right-hand twisted superstructures via solvent composition[124]. The degree of twist and pitch of ribbons could be limited to a one-handed helical structure by exploiting solvophobic effects. The measurement of circular dichroism confirmed its right-handed nature. The assembly of an aggregation-induced emission enhancement molecule is of use in chiral mechanochromic luminescent superstructure formation.

Two twisted donor–acceptor cruciform luminophores were prepared[125] which consisted of two π-conjugated segments, linked by a central benzene core. They exhibited intramolecular charge transfer emission, aggregation-induced emission, a high solid-state efficiency and – in particular – a substituent-dependent mechanofluorochromism. As-prepared powder samples of the more twisted luminophore emitted a strong blue-green light which was centered on 474nm. The fluorescence colour changed to yellowish-green (531nm) as a result of grinding, and this mechanochromism was reversed by fuming. A high-contrast behavior of one of the materials resulted from planarization of the molecule, and planar intramolecular charge-transfer under an external force. The less-twisted material exhibited no such mechanofluorochromic behavior because its conformation generated strong intermolecular forces and a resultant close-packing. Thus its crystalline powder-form had a higher lattice energy and superior structural stability.

A tetraphenylethene derivative containing a rhodamine unit has been synthesized[126] by using a high-efficiency Suzuki coupling reaction. The highly emissive target solid-state fluorescent molecule exhibited marked aggregation-induced emission enhancement, and also a reversible mechanochromic luminescence behavior, with a colour-change from orange to red. The latter effect was attributed to a morphological transformation between crystalline and amorphous states.

Two functional tetraphenylethylene derivatives, modified with vinylpyridine or vinylnitrobenzene, were synthesized[127] by using a Heck coupling reaction. They exhibited aggregation-induced emission activity in solution, and both exhibited reversible mechanochromism in the solid state. The effect of grinding was to cause a 13 to 40nm red-shift, in the fluorescence spectra, which could be reversed by solvent fuming. Three tetraphenylethene-based compounds[128] having differing substituents all exhibited aggregation-induced emission and mechanochromic luminescence behaviours. In one case, the mechanofluorochromic behavior was self-reversible. Two fluorescent coordination polymers have been synthesized[129] using a tetraphenylethene derivative. They exhibited a reversible mechanochromic luminescence with colour changes from blue to green-yellow, due to grinding, being visible to the unaided eye.

The material, (9,9-dimethyl-7-(4-(1,2,2-triphenylvinyl)phenyl) -9H-fluoren-2-yl) dimesitylborane, which comprises dimesitylboryl, fluorene and tetraphenylethene groups, exhibits aggregation-induced emission and a high thermal stability. Due to the dimesitylboryl group, it can interact with fluorine ions to cause a blue-shift in the absorption spectrum and an apparent decrease in the emission spectrum. It also has mechanochromic properties, and the solid-state fluorescence colour can be reversibly switched, between blue and sky-blue, by grinding and fuming[130].

The polymer, poly(9,10-anthracenevinylene-alt-4,4'-(9,9- bis(4-(4'-(1,2,2'-triphenylvinyl) phenoxy)butyl)-9H-fluorene-2,7-diyl) dibenzaldehyde, was synthesized[131] via the Wittig-Horner reaction, with fluorene and 9,10-distyrylanthracene as a backbone and tetraphenylethenes as pendant groups. It emits an orange light which is centered on 567nm, when in a dilute tetrahydrofuran solution, while the solid has a strong yellow emission centered on 541nm. The as-synthesized material exhibits aggregation-enhanced emission, a high thermal stability (decomposition temperature: 430C) and a high morphological stability (glass transition temperature: 171C). Under mechanical solicitation, the emission changed from yellow to red and this was attributed to destruction of the crystalline structure: conformational planarization of the distyrylanthracene components and an increase in the molecular conjugation of the backbone.

Two tetraphenylethene-based diphenylacrylonitrile derivatives exhibited[132] reversible mechanochromic luminescence and aggregation-induced emission, with a green fluorescence. This was changed to orange-yellow by grinding, but the green emission could be restored by fuming with dichloromethane. The reversible mechanochromism was attributed to transformation from a crystalline state to an amorphous state, and *vice versa*. Single-crystal studies of the structure revealed that poor intermolecular C-N (3.087Å) interactions led to a loose packing of the molecular conformation, thus facilitating mechanochromic behavior of the crystals during grinding. Tetraphenylethene conjugates were synthesized[133], yielding an aggregation-induced emissive fluorogen having mechanochromic properties: the emission changed from blue to green upon grinding, and from green to blue upon fuming. Investigation of the tetraphenylethene-naphthyridine structure revealed a high sensitivity to silver ions, relative to other metal ions, with a detection-limit of 0.25µM in aqueous solution. The stoichiometry of the complex comprising tetraphenylethene-naphthyridine and silver ions was 2:1.

Tetraphenylethene-substituted pyrazabole has been synthesized[134] by using the palladium-catalyzed Suzuki cross-coupling reaction. It exhibited marked aggregation-induced emission, and a reversible mechanochromism with a colour-change between blue and green. This was attributed to conversion of the crystalline state into an amorphous one. Two unsymmetrical tetraphenylethene substituted donor-acceptor benzothiadiazoles were similarly synthesized[135] by using the Suzuki cross-coupling reaction. The tetraphenylethene donor component was kept constant while the benzothiadiazole acceptor strength was modulated by using phenyl or cyanophenyl units. The products exhibited strong solvatochromic and aggregation-induced emission behaviors. The cyano-containing material exhibited a reversible mechanochromic behavior, with a marked colour change between green and yellow, while the other material exhibited no mechanochromism. The solid-state absorption and emission properties of both products exhibited differing behaviors in the pristine and ground conditions, and a reversible morphological change between crystalline and amorphous states occurred during grinding.

Tetraphenylethene-substituted phenanthroimidazoles were synthesized[136] by again using the Suzuki cross-coupling reaction. These materials exhibited a reversible mechanochromic behavior, with the colours changing between sky-blue and yellow-green. This was again attributed to conversion of a crystalline state into an amorphous state. Hydrogen-bonding interaction of a cyano-group in one of the materials resulted in enhanced aggregation-induced emission and an improved thermal stability.

A pyridinium-modified tetraphenylethene-based material was shown[137] to exhibit aggregation-induced emission enhancement and an irreversible mechanochromism. Tetraphenylethene substituted[138] with the electron-acceptor, 1,3-indandione, preserved the intrinsic aggregation-induced emission ability of tetraphenylethene, while manifesting intramolecular charge-transfer and a marked solvatochromic behavior. When the solvent was changed from apolar toluene to highly-polar acetonitrile, the emission peak suffered a red-shift from 543 to 597nm. Solid samples exhibited mechanochromism, in that grinding of as-prepared powder caused an emission red-shift from green (centered on 515nm) to orange (centered on 570nm). The mechanochromism was reversible by using grinding and annealing and grinding and solvent-fuming cycles. The emission of solid samples changes from orange, in the ground state, to yellow in the annealed or solvent-fumed states. The mechanochromism was attributed to a transition between amorphous and crystalline states. Because the material underwent hydrolysis in basic aqueous solutions, the red-orange emission could be quenched by OH⁻ or hydroxyl-generating species.

Five tetraphenylethene derivatives containing difluoroboron beta-diketonate complexes with various alkyl chain lengths have been synthesized[139]. The products exhibited strong solid-state fluorescence emissions, marked aggregation-induced emission and colour changes. They manifested an irregular mechanochromic behavior upon increasing the alkyl chain-length. Following grinding, the solid-state fluorescence was red-shifted and the degree of that shift was attributed mainly to the initial crystalline emission. The latter depended in turn upon the J-type molecular stacking and electron-density distribution. Ultrasonic treatment led to a blue-shift of the solid-state fluorescence. The inherent mechanochromism mechanism was attributed to the collapse of J-type stacking under ultrasonic treatment.

Tetraphenylethylene is assumed to exhibit aggregation-induced emission enhancement due partly to a restriction of intramolecular rotation, and some halogen-substituted fluorophores have interesting luminescent properties due to C-H···Halogen interactions. By modifying tetraphenylethylene with a difluoroboron β-diketonate fluorophore, three complexes were synthesized[140], all of which exhibited intramolecular charge-transfer, a high fluorescence quantum yield, aggregation-induced emission enhancement and mechanochromism. By using grinding and fuming or grinding and annealing cycles, the original emission colour of each material could be almost completely recovered. One material exhibited spontaneous reversibility following grinding at room temperature. Due to the effect of halogens such as chlorine and bromine, two of the materials exhibited a polymorphism-dependent fluorescence with yellow plate-like crystals and green needle-like crystals before and after grinding. The maximum fluorescence emission wavelength of these materials was shifted bathochromically by 70 or 59nm due to grinding. The mechanochromism of these complexes was related to transformations between amorphous and crystalline states and to metastable-state formation due to grinding, rather than to molecular planarization, excimer or exciplex formation.

Tetraphenylethylene-based dumbbell-shaped molecules were synthesized[141] so as to contain alkyl chains with 7, 8, 9 or 10 carbon atoms. The materials exhibited aggregation-induced emission behaviour, and the effect of the alkyl chain-length was investigated by studying the morphology of self-assembled nanostructures which formed in various tetrahydrofuran/water solvents. At a water fraction of 80%, tetraphenylethylene derivatives with odd alkyl chains self-assembled into nanosphere structures, while those with 8 alkyl chains formed microbelts and those with 10 alkyl chains aggregated into flower-like superstructures. These derivatives exhibited mechanochromism upon grinding, indicating the importance of molecular stacking to luminescent properties.

A similar synthesis[142] of amphiphilic and dumb-bell shaped aggregation-induced emission tetraphenylethylene derivatives showed that both of them formed three-dimensional flower-shaped supramolecular structure in tetrahydrofuran/water solutions at various water fractions. The derivatives also exhibited mechanochromism during grinding. Self-assembly of the derivatives was attributed to two principal factors: the tetraphenylethylene core and alkyl chains which optimized dispersive interactions, and amide-linkage via molecular recognition.

Water-soluble tetraphenylethylene, containing four sulfonate groups as a sodium salt, was investigated[143] with regard to its aggregation-induced emission behaviour by an adding organic solvent to an aqueous solution. The product was weakly emissive in pure water, but emitted strongly upon adding tetrahydrofuran. The emission properties and aggregate morphologies depended markedly upon the solution pH. Uniform nanorods with a width of about 200nm and a length of up to 10μm were obtained at solution pH of 1. The materials also exhibited very good mechanochromic properties.

Donor-acceptor-donor organic boron complexes which contained tetraphenylethylene and various aromatic amine units have been synthesized[144]. They exhibited aggregation-quenching, aggregation-induced emission and dual-state emission fluorescence. The orange crystals of one of the compounds had a twisted molecular configuration and a head-to-tail J-type stacking, while the red crystals of another compound had sparse super-unit cells with two face-to-face anti-parallel dimers and four other single molecules. One anti-parallel head-to-head J-type stacking led to π-π stacking interactions. A faint fluorescence emission from one material was attributed to parallel J-type stacking. The materials also exhibited reversible multi-colour switching, blue-shifted fluorescence emission and a notable luminescence enhancement. It was concluded that the molecular stacking-mode was the critical factor determining the optical and mechanochromic properties.

Two aryl-substituted tetraphenylfurans and corresponding ring-opened (Z)-1,4-enedione derivatives have been synthesized[145] which had propeller-like configurations. The tetraphenylfurans were aggregation-quenched fluorophores while the 1,4-enediones were aggregation-induced emissive. The tetraphenylfurans and enediones both exhibited a so-called turn-on mechanochromism, but with a blue-shift of the mechanochromic fluorescence for the tetraphenylfurans and a red-shifted mechanochromism for the enediones. Both optical phenomena were very dependent upon the type of molecular packing.

Mechanochromism   Materials Research Forum LLC
Materials Research Foundations **52** (2019)   https://doi.org/10.21741/9781644900277

Six tetraphenylpyrazine-based luminogens were designed[146] so as to contain various conjugation and donor-acceptor components. The materials exhibited bright-blue to red emissions in film form, and the photo-physical behaviour was attributed to a competition between a restriction of intramolecular motion and intramolecular charge-transfer. One of the materials could be used as a reversible mechanochromic material.

Silole derivatives are silicon-containing conjugated rings with $\sigma^*-\pi^*$ conjugation, but are not good blue luminogens due to their relatively long conjugate lengths. By replacing silicon with carbon, six new aggregation-induced luminogens without any $\sigma^*-\pi^*$ conjugation effect were synthesized[147] on the basis of a tetraphenylcyclopentadiene core. In addition to a sky-blue (492nm) emission, control of the degree of conjugation by changing the linkage mode produced a blue (440nm) emission. There was also a mechanochromism in which the fluorescence changed between deep-blue and green.

By building on a skeleton of tetraphenylpyrazine, an aggregation-induced emitter was produced[148] via the addition of a malonitrile group. It exhibited an abnormal reversible mechanochromism, with hypsochromic effect. A mechanochromic material was obtained[149] by attaching the boron atom of BH_3 to the amine link between tetraphenylethylene and rhodamine-B units. The product exhibited the novel ability to switch colours from dark-blue to bluish-green, and then to reddish, during grinding.

Donor-acceptor benzothiazole-substituted tetraphenylethylenes were synthesized[150] in order to investigate the influence, of the linkage between the two components, upon the mechanochromic properties. The aggregation-induced emissions and mechanochromism in fact depended upon the linkage-point (ortho, meta or para). The meta-isomer exhibited the greatest (51nm) grinding-induced shift while the ortho-isomer exhibited the smallest (9nm) one. Single-crystal X-ray studies revealed the presence of a highly twisted conformation and tight-packing in the ortho-isomer as compared to that in the meta-isomer.

4,4'-((1Z,3E)-1,2,3,4-tetraphenylbuta-1,3-diene-1,4-diyl)-dibenzaldehyde exhibits[151] a blue-shift of 23nm during grinding which cannot be restored by fuming with organic solvents or by heating. On the other hand, due to its twisted conformation and relatively loose packing, 4,4'-((1Z,3Z)-1,2,3,4-tetra-phenylbuta-1,3-diene-1,4-diyl)dibenzaldehyde exhibited a reversible red-shift of 20nm during repeated cycles of grinding and fuming.

Three D-π-A type quinoxalines were modified[152] with tetraphenylethylenes. Tetraphenylethylene substitution at the 5,8-positions of the quinoxaline led to absorption bands at 316nm and 303nm which originated from π-π^* transitions. When tetraphenylethylene units were located at the 2,3-positions of quinoxaline, the π-π^* transition absorption was blue-shifted to 287nm because of poor planarity and weak conjugation. All three compounds exhibited aggregation-induced emission. The solid-state mechanochromic emission colours of two of the materials were very different before and after grinding. As-prepared crystals of one compound emitted blue light under ultra-violet irradiation, but these could be changed into an amorphous powder emitting bluish-green light. The mechanochromism was attributed to a transition between the crystalline and amorphous states, and was reversible by heating or fuming.

Molecules of the form, D-A-A′, where A′ was benzothiazole, A was benzothiadiazole and D was tetraphenylethylene, were synthesized[153], with para-, meta- and ortho-attachments of tetraphenylethylene to the benzothiazole. The position of the attachment affected the acceptor strength and molecular packing. A reversible mechanochromic behaviour was associated with the transition from crystalline to amorphous. Single-crystal data showed that the mechanochromism depended upon flexing and twisting of the donor and acceptor units.

A material which contained pyrene and rhodamine-B chromophores[154], separated by a peptide spacer, exhibited[155] a fluorescence which could be changed from blue to bluish-green and reddish by grinding[156]. The tetraphenylalanine spacer was judged to play a crucial role in forming the original blue powder. An emission centered at 440nm from the initial powder was red-shifted to 480nm by grinding. Following further mechanical treatment, a new peak at 583nm was detected from the reddish powder. An emission again appeared at 440nm when the reddish powder was heated and treated with a solvent. The reddish color was attributed to a force-induced chemical transformation of rhodamine-B from a spirolactam to a ring-opened amide.

In recent work, a tetraphenylethene derivative, with substituted sulfonyl-based naphthalimide units, was synthesized[157]. The target molecule exhibited aggregation-induced emissions, but the linkage between the tetraphenylethene and naphthalimide units was non-conjugated. The material also exhibited a surprising and highly-reversible mechanochromism in the solid state. This was attributed to a change in the form of aggregation occurring between the crystalline and amorphous states. In other studies[158], the Suzuki-Miyaura cross-coupling reaction of bromo-alkenes with fluorophenylboronic acid yielded 17 fluorinated tetraphenylethene compounds which contained fluorine substituents alone directly on the tetraphenylethene core, with various numbers and substitution points of fluorine atoms. Four of the derivatives had C-H⋯F and C-H⋯π hydrogen bonds in the crystal structure. These exhibited aggregation-induced emission characteristics, with the aggregation step being gradual and time-dependent.

A multi-responsive fluorescent tetraphenylethylene derivative has recently been synthesized[159] which comprises a tetraphenylethylene skeleton having two methoxyl and one carboxyl peripheral groups. The material exhibits a typical aggregation-induced emission behavior, plus a multi-coloured mechanochromism. The emission can be reversibly switched between blue (462nm), bright cyan (482nm) and yellow (496nm); with a high contrast for solvent fuming, heating or grinding. The molecule can occur in 4 crystalline states, following crystallization from different solvents. The multi-coloured mechanochromism was related to the various interactions and packing-modes of the molecules within the crystals.

Diketones

Difluoroboron β-diketonate complexes exhibit a strong fluorescence, both in solution and in the solid state, with large extinction coefficients and tunable emissions. Some of them exhibit mechanochromism, and a room-temperature phosphorescence in the solid state.

Mechanochromic luminescence, with a hypochromic shift of the emission due to grinding, has been studied[160] in carbazole-containing difluoroboron β-diketonate complexes. These exhibit strong fluorescence in both the solid state and in organic solvents. The emission of structurally symmetrical complexes is influenced less by the polarity of the solvent than is that of asymmetrical complexes, and they maintain a strong fluorescence in highly polar solvents. The good planarity of the carbazole group and the intramolecular charge transfer behaviour of the complexes cause them to exhibit two-photon excited fluorescence. In the solid state, the emission colours can be tuned reversibly by external influences and they exhibit an unusual hypochromic shift due to grinding. Hypsochromic and bathochromic shifts of the emission can be produced by differing sample treatments. The initial close-packed molecular structure is deemed to be responsible for a blue-shifted emission.

Because of their easy synthesis, β-diketones are convenient bases for the preparation of mechanochromic luminescent materials, but cannot yet offer a full range (red, green, blue) of luminescent colours. One attempt to prepare such β-diketone emitters with a red-shifted emission involved the substitution of various donor and acceptor groups and coordination of BF_2 to the β-diketonate. In another attempt[161], 3,4,5-trimethoxy-substituted phenyl rings were used to disrupt the molecular packing so as to produce mechanochromic luminescence and aggregation-induced emission. Trimethoxy-substituted β-diketones were prepared via Claisen condensation, and para-substituted with fluorine, iodine or cyanogen groups. Boronated complexes were then synthesized by reaction with boron trifluoride diethyl etherate. Density functional theory calculations indicated that the para-substituted methoxy groups were directed out of the molecular plane, thus possibly loosening the solid-state packing. Aggregation-induced emission

Materials Research Forum LLC

https://doi.org/10.21741/9781644900277

studies, using tetrahydrofuran/H_2O solutions, revealed the presence of aggregation-induced emission in β-diketone dyes and in boron complexes. High-contrast mechanochromic luminescence was observed for each dye, and this dye-set encompassed the entire visible spectrum.

Difluoroboron β-diketonate compounds exhibit a solid-state switchable luminescence which tends to be seen as a blue-shifted emission when dye films are annealed, and which is followed by a red-shift under mechanical shearing. Methoxy-substituted dyes did not exhibit any thermal or mechano-responsive behavior, but the addition of longer alkoxyl-chain substituents led to a stimulus-responsive behavior. Furan- and thiophene dodecyloxy-substituted dyes exhibited reversible luminescence switching between crystalline blue-shifted, and amorphous red-shifted, states[162]. Gentle heating of the furan dye produced a green emission while melting, followed by rapid cooling, produced an orange form. The thiophene dye exhibited a blue-shifted emission when annealed below its melting point and a red-shifted emission when mechanically treated. These transformations were completely reversible in both dyes.

Substitution at the α-position of difluoroboron β-diketonates can force these dyes out of their planar conformation and produce an aggregation-induced enhanced emission behaviour. If mechanochromic luminescence truly arises from the postulated formation of H-aggregates during shearing, substitution at the α-position could then hinder the process and affect the mechanochromism. In order to test this, luminescent difluoroboron β-diketonate derivatives, with substituents at the α-position, were synthesized[163]. A difluoroboron dibenzoylethane derivative, with a methyl group at the α-position, exhibited a bright blue emission in the solid state but was essentially non-emitting in solution. It also exhibited clear aggregation-induced enhanced emission in various water fractions, with dimethyl sulfoxide or tetrahydrofuran as the solvent. A difluoroboron 2-benzoyl-1-tetralone derivative was much brighter in solution and a decrease in emission intensity was observed upon inducing aggregation. A methoxyphenyl-substituted dye exhibited bright emission from a twisted intramolecular charge transfer state in solution and an attenuated emission intensity in low water-fractions and an aggregation-induced enhanced emission at higher water fractions with dimethyl sulfoxide as the solvent.

Difluoroboron β-diketonate compounds exhibit solid-state luminescence phenomena which can be tuned by altering the molecular structure so that dyes with various halide substituents exhibit particular mechanochromic luminescence, mechanochromic luminescence quenching and solid-state emission behaviours. A series of difluoroboron dibenzoylmethane dyes have been synthesized[164], with CH_3, C_5H_{11}, C_6H_{13}, $C_{12}H_{25}$ or $C_{18}H_{37}$ additions. By keeping to the same heavy atom, the dependence of the mechanochromic luminescence upon alkyl chain-length could be isolated. The hydrogen derivative was only weakly emissive in the solid state, and exhibited only a minimal mechano-responsive behavior. Alkoxy dyes exhibited a tunable mechanochromic luminescence and mechanochromic luminescence quenching which depended upon the length of the alkyl chain. Longer chains corresponded to smaller singlet triplet energy gaps, greater triplet emission enhancement (77K), and slower recovery following deformation under ambient conditions. Shorter chains had a much greater tendency to produce ordered emissive states. Single-crystal X-ray diffraction studies revealed appreciable differences in crystal-packing and π-π stacked dimers between dyes with alkyl chains and those without alkyl chains.

Metal-free methoxy-substituted dinaphthoylmethane β-diketone exhibits an aggregation-induced emission and mechanochromic luminescence that recover rapidly at room temperature. The effect of substituents and boron coordination has been investigated by preparing[165] a series of methoxy- and bromo-substituted derivatives and their corresponding boron complexes. All of the boron complexes exhibited a red-shifted absorption and emission, as well as larger solution and solid-state quantum yields, than did β-diketones. Aggregation-induced emission studies revealed increased emissions in the case of methoxy and brominated methoxy derivatives, but the emission of the corresponding boron complexes was diminished by aggregation. The boron complexes remained strongly emissive in the solid state. The mechanochromic luminescence properties were investigated by using spin-cast films, showing that deformation resulted in the appearance of a blue-green emission in ligands and a colour-change from green to yellow-orange in boron complexes. The bromide-substituted derivatives exhibited increased room-temperature recovery times, as compared with other dinaphthoylmethane ligands, while boron complexes exhibited only a partial recovery after several days.

Mechanochromic luminescence occurs in methoxy-substituted dinaphthoylmethane ligands, even without any coordination to boron[166]. The material also exhibits aggregation-induced emission. Diketone thin films exhibited high-contrast emission when deformed, with rapid recovery of the pre-deformed state at room temperature.

Difluoroboron β-diketonate compounds with substituted hydrogen, fluorine, chlorine, bromine and iodine were synthesized[167], and their luminescence properties studied in CH_2Cl_2 solution, as spin-cast films on glass substrates or as bulk powders. All of the derivatives exhibited mechanochromic luminescence and mechanochromic luminescence quenching. The solid-state emission was affected by the annealing temperature: fluorine, chlorine and bromine products required higher annealing temperatures than did hydrogen- or iodine-substituted analogues in order to obtain ordered emissive states.

Spin-cast films on glass underwent a structural change from amorphous, in the as-spun state, to an ordered crystalline after annealing. The fluorine and chlorine derivatives had a tendency to form large lamellar crystallites. Halogen-substitution markedly impaired the ability of the material to recover its ordered emissive state following deformation under ambient conditions.

Difluoroboron β-diketonate dyes with methyl, phenyl, naphthyl and anthracyl groups have been used[168] to deduce the effect of π-conjugation length and arene size on mechanochromic luminescence in powder and spin-cast film samples. These materials emitted at wavelengths ranging from blue to red; depending upon the above conjugation length. All of the derivatives, apart from the methyl-phenyl one, exhibited an emission change following mechanical treatment, with increasing π-conjugation being related to a greater red-shift in fluorescence. Recovery was greatly affected by the nature of the aromatic substituent.

In order to determine the effect of alkyl chain-length upon solid-state photoluminescence and mechanochromic luminescence, compounds with dibenzoylmethane ligands and alkoxyl substituents were prepared[169]. The additions were of the form, C_nH_{2n+1}, where n = 1, 2, 3, 5, 6, 12, 14, 16 or 18. The fluorescence spectra and lifetimes were essentially identical in CH_2Cl_2 solution, but differed from sample to sample when the latter were powders or films. The recovery-time tended to increase with alkyl chain-length, and ranged from minutes for n = 3 to days for n = 18. Chains with n = 6, 12, 14, 16 or 18 did not fully return to the pre-deformation emissive state, even after months. Difluoroboron dibenzoylmethane-polylactide polymers similarly exhibit[170] a molecular-weight dependent fluorescence in where the polymer chain plays an important role. When a lipid chain was used to replace the polylactide, a reversible mechanochromic fluorescence was observed. Annealed spin-cast films of the material on glass emitted a blue fluorescence under ultra-violet light but, following abrasion, the mechanically disturbed regions turned yellowish-green. The blue colouration could be recovered by heat-treatment. It was concluded that ordered-to-amorphous structural changes which occurred during mechanical treatment could increase molecular rotational freedom and thus permit more efficient excimer emission; which would tend to occur at longer wavelengths.

Three different states were identified in difluoroboron-β-diketones[171]: two crystalline ones (stable green-emissive, metastable yellow-emissive) and an amorphous one (yellow-orange). Dynamic excimer formation could occur in the yellow-emissive crystalline and amorphous states. Switching between those states during mechanical treatments led to the observed fluorescence colour-changes.

Thin films of a methoxy-substituted β-diketone initially exhibited[172] a blue (428nm) emission, but a green (478nm) emission was found following deformation. This mechanically-generated state recovered extremely rapidly. The recovery was largely substrate-independent, but film thickness was thought to be the main factor governing the recovery. The β-diketones exhibit high-contrast mechanoluminescence and aggregation-induced emission, but tend to emit over a narrow range (~420 to 500nm). In order to widen this luminescence colour-range, dimethylamino-substituted β-diketones have been synthesized[173]. Most of the products were responsive to mechanical stimuli. The solid-state emission wavelengths were related to the electron-withdrawing power of the para-substituent and ranged from blue to orange (488 to 578nm). These changes were the result of crystalline-to-amorphous phase transitions.

A boron difluoride quinoline β-diketone which existed as various polymorphs was synthesized[174] and the various structures were examined. Each polymorph had a different stacking orientation of the chromophore components, and involved several different intermolecular interactions; all of which affected their emissions. The molecule was mechanochromic, with the fluorescence of solid powder and crystals being reversibly affected by mechanical forces. The three types of polymorph had differing emissions. One crystalline sample exhibited a yellow emission (567nm), similar to that (566nm) of solid powder. Rod-like and tetragonal crystals emitted at 581 and 601nm, respectively. These variations in emission were attributed to their differing molecular packings. Grinding one of the crystals changed its emission from orange to red (598nm). Conversion between monomer and π-π stacking-induced aggregates was suggested to be the mechanochromism mechanism.

In the latest work on these materials[175], β-diketones and the corresponding mono-substituted, meta-position di-substituted and tri-substituted boron complexes were extensively studied. Experiment and prediction both showed that the mechanical responses of the 6 molecules were dependent upon the structure, and were affected mainly by conjugation and steric effects arising from the various substitutions. The di-substituted materials exhibited the most marked mechanoluminescent behavior whereas mono-substituted materials exhibited essentially no response to mechanical solicitation. The tri-substituted compounds exhibited fast self-erasure properties which might be of potential use in optical force sensors.

Just recently, a series of asymmetrical diphenylketone derivatives possessing high-contrast mechanochromic properties has been discovered[176]. The members exhibit dual-emission bands by combining a traditional fluorescence (426 to 459nm) with intermolecular thermally-activated delayed fluorescence (565 to 575nm), and the solid-state high-contrast mechanochromism goes from blue and cold-white to yellow.

Anthracenes

It will be obvious from other sections, such as the one dealing with spiropyran, that anthracene is an important component of many mechanochromic molecules. It will be considered in its own right here. A uni-chromophoric anthracene-pentiptycene derivative has been found[177] to exhibit both mechanofluorochromism and photo-mechanofluorochromism. This leads to the appearance of various coloured emissions, including red-green-blue and almost pure white. One compound crystallizes as two polymorphs: a yellow and a green emitter. Pairwise-stacked anthracene groups here undergo photo-dimerization to form a black (ultra-violet) photo-dimer emitter and exert photo-mechanical stresses on neighbouring molecules. Such photo-mechanical stresses cause excimer-to-monomer switching and result in a blue fluorescence of the yellow form. A red-emissive so-called super-dimer is generated in the case of the green form. Recovery of the yellow form requires annealing, while the green form can be restored by selective photo-excitation of the super dimer.

The mechanochromic luminescence of 9,10-distyrylanthracene derivatives has been investigated[178] by subjecting single crystals to hydrostatic pressure in a diamond-anvil cell, and comparing the effects with those of grinding. During such grinding, three single crystals exhibited blue-shifts of the fluorescent spectra due to conversion of the ordered structure into an amorphous state, but almost invisible colour changes. The high-pressure results revealed that three 9,10-distyrylanthracene forms all underwent large red-shifts. High-pressure absorption spectra showed that a new charge-transfer state appeared when -CH_3 or -2CN substituents were introduced into the 9,10-distyrylanthracene; and underwent clear red-shifts. When the pressure was relaxed back to ambient, the fluorescence and absorption spectra almost completely recovered, thus suggesting that the structural changes were reversible within a certain pressure region. The donor–acceptor system, 4-(anthracen-9-yl)-N,N-diphenylaniline, has similarly been subjected[179] to high-pressure Raman and fluorescence experiments in the form of ordered crystalline powder. The spectroscopic features seen under grinding shear forces were compared with those produced by hydrostatic pressures in a diamond anvil cell. During that compression, the crystals underwent a clear emission-band shift from 476 to 600nm, together with an intramolecular charge-transfer which was independent of any hybridized local or other charge-transfer and which was very sensitive to external forces. When the pressure was relaxed, the Raman and fluorescence spectra both completely recovered.

A so-called butterfly molecule, 9,10-bis(2-phenyl-2-(pyridin-2-yl)vinyl)anthracene, has been studied[180] in which multiple rotatable aryl groups permit the material to exhibit a marked aggregation-induced emission effect, with a highly twisted butterfly-like backbone where intermolecular π-π interactions help to promote mechanofluorochromism. Theory suggests that protonation of the pyridyl units can narrow the band-gap, while X-ray crystallography shows that the molecules are packed into a Z-configuration where the pyridyl units adopt a more twisted conformation, with isolated N-electron pairs being stabilized by intra- and inter-molecular CH···N and CH···π interactions. In previous work, bis(diarylmethylene)dihydroanthracenes having butterfly-shapes had been studied[181] which exhibited aggregation-induced emission due to a restriction of intramolecular motion in the aggregated state. They displayed mechanochromism because of transformations between crystalline and amorphous states, aided by grinding or solvent fuming, which yielded various colours.

Experimental molecules have been synthesized[182], using Schiff-base methods, which have rhodamine-B and anthracene as building units, plus various spacers. These small structural variations lead to marked differences in the mechanochromic properties of solid-state samples. Both undergo a force-induced colour change, and one of them exhibits two colour-changes due to ring-opening isomerization of the rhodamine derivatives. The other material exhibits a three-colour change under the action of a force, and this is attributed to transformation of the various packing-modes of anthracene and to force-driven chemical reaction of the rhodamine-B. In previous work[183], force-induced reversible three-colour switching had been observed in molecules which comprised an anthracene unit and a rhodamine-6G component which were bridged by an oligopeptide of tetraphenylalanine. The multi-colour switching was attributed to a synergetic effect involving supermolecular-structure change and chemical-structure alteration. Various π–π overlapping interactions of neighboring anthracene rings, and the ring-opening reaction of rhodamine-6G, led to switching between bluish-green, blue and yellow. In a molecule having two anthracene components, and a rhodamine-6G unit on both sides of a branched amidoamine spacer, the emission properties had already[184] been attributed to π-π overlap of the anthracene components and to the ring-opening reaction of rhodamine-6G.

An anthracene-based compound, 9,10-bis(4,5-diphenyl)-imidazol2-yl)-anthracene, has been synthesized[185] which has a rigid conjugated structure and which exhibits a strong blue fluorescence in the crystalline state. Grinding its crystals changes the original blue emission to green. The blue colour can be recovered by fuming with isobutanol, and the overall behaviour is attributed to changes in the π-π stacking between aromatic rings during mechanical treatment and fuming.

A study has been made[186] of the spontaneous recovery of the mechanochromic luminescence of carborane-anthracene derivatives. Recovery was attributed to metastable charge-transfer emission from anthracene to ortho-carborane. The particular structure of the ortho form was essential to obtaining the required properties.

Samples of 10,10′-bis(2-(N-alkylphenothiazine-3-yl)vinyl)-9,9′-bianthracene, with alkyl chain lengths of 2, 8, 12 or 16, were synthesized[187] in order to determine the effect of chain-length upon solid-state fluorescence. The compounds emitted a strong fluorescence in solution and in the solid state. The fluorescence emission and grinding-induced spectral shifts of the solid were alkyl-length dependent. Samples with a chain-length of 2 exhibited the weakest fluorescence, and the absorption-spectrum was shifted by mechanical force. Samples with longer alkyl chains manifested a similar mechanochromism and a greater fluorescence contrast following grinding. The fluorescence of ground solid samples with a chain-length of 16 recovered at room temperature, while other compounds required higher temperatures. The cold-crystallization temperature-difference was deemed to be responsible for thermal restoration.

An investigation was made of the optical properties and stimulus-response of 2,6-bis(diethoxylphosphorylmethyl)-9,10-bis(N-alkylcarbazol-3-yl-vinyl)-anthracene[188]. This exhibited an aggregation-induced emission behaviour and reversible mechanochromic luminescence. As-received powders, formed by rotary evaporation or recrystallization were green-emitting (510nm). Grinding revealed that materials with branched 2-ethylhexyl components were all mechanofluorochromic and underwent a grinding-induced red-shift. When the ground samples were annealed and then cooled to room temperature, or were exposed to dichloromethane vapour at room temperature, the fluorescence reverted to the as-received state. This grinding-fuming cycle could be repeated.

A molecule (figure 12) was synthesized[189] from aspartic acid and phenylalanine, together with an anthracene group. Its original blue emission changed to green following gentle grinding. The emission of the as-received sample was centered on 445nm, with a shoulder peak at 460nm. Following grinding, the band became broader and less-structured and was red-shifted to 468nm, with a shoulder peak at 445nm. The blue emission could be recovered by heating (150C, 600s) and this cycle could be repeated several times, although the intensity decreased. The differing colours of the as-received and ground samples were attributed to an overlapping π-π packing of the anthracene units. The emission colours of anthracene greatly depend upon the degree of π-π interaction: a crystal with strong interchromophore π-π interactions give a green colour whereas weak π-π interactions result in emissions of sky-blue or deep-blue. Similarly, the blue colour of the present as-received material was attributed to the less overlapped aromatic units, while the green color following grinding suggested that there was increased π-π interaction of the anthracene groups when a force was applied. The C=O stretching band shifted from 1633 to 1638/cm and the N–H band shifted from 1526 to 1523/cm; indicated a partial disruption of the hydrogen bonds which might lead to modification of the molecular arrangements.

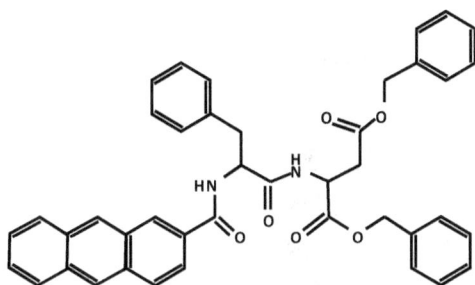

Figure 12. Structure of molecule based upon aspartic acid and phenylalanine with an anthracene group

Acetylenes

Donor-π-acceptor compounds have been studied[190] which comprised diphenyl aminophenylacetylene as a donor-π component and a hetero-aromatic ring-bearing ester as an acceptor. Those compounds incorporating a dicyanobenzoic ester underwent a bathochromic shift during grinding, while those incorporating an ester with benzene, imidazole and thiazole rings underwent a hypsochromic shift. The compounds which exhibited bathochromic shifts had a herring-bone alignment, with hydrogen-aggregate like π-stacking of the crystal structure, whereas those which exhibited hypsochromic shifts had structures featuring an alignment of the long axes of the molecule.

Azoles

A donor-acceptor-donor structured conjugate, 4,7-di(2-thienyl)-2,1,3-benzothiadiazole, comprising benzobis(1,2,5-thiadiazole) and thiophene units, has been synthesized[191]. It manifested intramolecular charge transfer, crystallization-induced phosphorescence and a marked mechanochromism in that crystals exhibited a bright-red (616nm) room-temperature phosphorescence which changed to a yellow (578nm) excimer fluorescence following grinding.

Two bipolar blue molecules, 2-(4-(4,5-diphenyl-2-(4-(1,2,2-triphenylvinyl)phenyl)-1H-imidazol-1-yl)phenyl)-1-phenyl-1H-phenanthro[9,10-d]imidazole and 1-phenyl-2-(4-(2-(4-(1,2,2-triphenylvinyl)phenyl)-1H-phenanthro[9,10-d]imidazol-1-yl)phenyl)-1H-phenanthro[9,10-d]imidazole, have been synthesized[192] which exhibit aggregation-induced emission and reversible mechanochromism. Pristine solid powder samples were white, and emitted blue light. Following grinding, the solid became yellow and emitted blue-green light. These colour-changes were visible to the unaided eye and could be reversed by solvent fuming.

Table 3. Colour changes of ground 4-(1H-indol-2-yl)-2,1,3-benzothiadiazole derivatives

R^1	R^2	Condition	λ(nm)
Me	Boc	as-received	485
Me	Boc	ground	525
Et	Boc	as-received	482
Et	Boc	ground	518
Pr	Boc	as-received	484
Pr	Boc	ground	518
Me	EtOCO	as-received	493
Me	EtOCO	ground	525
Me	Ts	as-received	483
Me	Ts	ground	517
Me	PhCO	as-received	506
Me	PhCO	ground	544
Me	BuCO	as-received	543
Me	BuCO	ground	564
Me	Me	as-received	580
Me	Me	ground	600

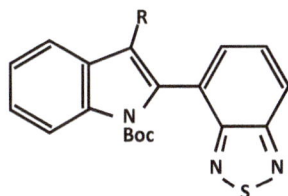

Figure 13.. Structures of 4-(1H-indol-2-yl)-2,1,3-benzothiadiazole derivatives

The mechanochromic luminescence properties of 4-(1H-indol-2-yl)-2,1,3-benzothiadiazole could be modified[193] by introducing various substituents (figure 13) into the indole ring (table 3). X-ray diffraction studies revealed either slipped stacks, or no intermolecular stacking, of the benzothiadiazole rings in the crystalline structure. The fluorescence spectra following mechanical grinding were similar to those of the molten fluorophores. Amorphization of the crystalline fluorophores was therefore judged to be responsible for the change in emission-colour (figure 14) during mechanical treatment, and their recovery was attributed to spontaneous recrystallization of the amorphous fluorophores.

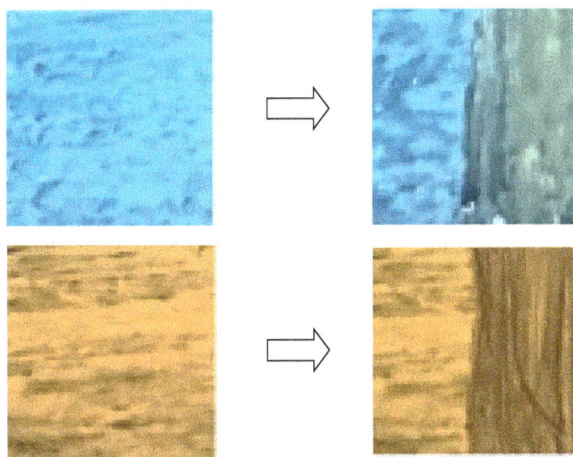

Figure 14. Examples of colour-changes
Upper: R^1 = Me, R^2 = Boc; lower: R^1 = Me, R^2 = Me

Five benzimidazole-based derivatives, in the form of D-π-A structures with bulky aromatic groups that could undergo twisted molecular conformation, were studied[194] showing that the presence of the –OH group could cause excited-state intramolecular proton transfer while various substituents on the π-conjugated backbone could vary the solid-state emission from blue to red. The solid-state emissions were increased to various degrees by aggregation-induced emission. The spectral properties of solid-state fluorophores were sensitive to mechanical stimulus. A reversible mechanochromic luminescence, using grinding and fuming, was observed in most of the materials.

Four carbazole derivatives having a $N(CH_3)_2$, H, Br or CN terminal group have been studied[195], showing that their twisted structures, intermolecular interactions and loose molecular packing are functions of the terminal group as are the aggregation-induced behaviors and mechanochromic properties. The $N(CH_3)_2$ derivative exhibited an aggregation-related quenching due to strong π-π intermolecular interaction, while the other derivatives exhibited marked aggregation-enhanced emissions. The differing twisted conformations and molecular arrangements led to various green-yellow to red mechanochromic responses among the derivatives, with red-shifts of 63, 30, 56 and 49nm, respectively. The H derivative was the only one to exhibit an irreversible mechanochromism.

A difluorobenzothiadiazole-based fluorescent material with a D-π-A-π-D structure exhibited[196] a reversible solid-state mechanofluorochromic behaviour, with its red fluorescent emission switching to near-infrared fluorescence during mechanical treatment. It recovered following fuming of the ground solid powder with dichloromethane.

The solid-state emission of N-Boc-indolylbenzothiadiazoles covers a wide fraction of the visible spectrum, and the colour can be easily varied[197] by changing the substituents on the two hetero-aromatic rings with 3-methylindolyl derivatives in particular imparting a self-recovering mechanochromic luminescence. That is, following mechanical treatment, the original solid-state emission is recovered spontaneously at room temperature. The recovery-time as well as the emission colour could be varied by introducing various components into the benzothiadiazole ring. The self-recovery of 3-methylindolylbenzothiadiazole compounds was attributed to partial amorphization of the crystals by mechanical treatment and its recovery via recrystallization.

Three boron 2-(2′-pyridyl)imidazole complexes having various different aromatic side-groups have been studied[198], showing that the solid-state emission and mechanochromic behaviors were a function of the type of side-group. A high colour contrast and a reversible mechanochromism were observed in all three analogues, with that isomer containing the more bulky thienothiophene group exhibiting a more marked mechanochromic behaviour. The steric and electronic properties of the aromatic substituents were shown to play an important role in controlling intermolecular interactions and intramolecular charge transfer. Another series of boron 2-(2'-pyridyl)imidazole complexes was found[199] to exhibit mechanochromic properties, upon which the terminal substituent had a large effect. When the basic material contained dimethylamino units which could freely rotate, it emitted poorly in most organic solvents, and as an amorphous powder, while a visible emission was observed in the crystalline state. Materials which comprised cyclohexylamino and diphenylamino units exhibited higher fluorescence efficiencies in solution and in the solid state, due to a greater steric hindrance, but mechanochromism was found only for materials containing dimethylamino and diphenylamino units. A change from crystalline to amorphous states during grinding was deduced to be the driving force for mechanochromism, and the flexibility of cyclohexylamino groups tended to make the material more amorphous, and thus non-responsive to mechanical disruption.

Three positional isomers (ortho, meta, para) of phenanthro-imidazoles were studied[200] with regard to aggregation-induced emission and mechanochromism. They exhibited a marked aggregation-induced emission, together with a reversible mechanochromism involving contrasting blue and green colours. The ortho and meta isomers underwent a grinding-produced spectral shift of 98nm while the para isomer exhibited a spectral shift of 43nm. In a previous study, four phenanthro-imidazoles had been synthesized[201], revealing aggregation-induced emission and a reversible mechanochromism between blue and green. The solid-state mechanochromic behavior of tetraphenylethylene-substituted phenanthroimidazoles was a function of the end-group of the phenanthroimidazole.

Two blue fluorescent materials having an asymmetrical structure, 2-(4'-((4-(9H-carbazol-9-yl)phenyl)sulfonyl)-[1,1'-biphenyl]-4-yl)-1-(4-(tert-butyl)phenyl)-1H-phenanthro[9,10-d]imidazole and 2-(4'-((4'-(9H-carbazol-9-yl)-[1,1'-biphenyl]-4-yl)sulfonyl)-[1,1'-biphenyl]-4-yl)-1-(4-(tert-butyl)phenyl)-1H-phenanthro[9,10-d]imidazole, which consisted of a sulfone group as the electron acceptor and carbazole or phenanthroimidazole as electron donors were synthesized[202]. Both compounds exhibited a mechanochromic behaviour in which the emission changed from deep-blue to blue-green during mechanical treatment; the emissions red-shifting by 50nm and 23nm, respectively, during grinding.

Two pyrene-based solid-state pyrenoimidazoles were synthesized[203], one of which had a twisted conformation. Both materials exhibited a marked aggregation-induced emission. with a reversible mechanochromic colour-change between blue and green.

Carbazole-based triphenylacrylonitrile derivatives were prepared[204] which exhibited typical aggregation-induced emission characteristics, and one derivative also exhibited mechanochromic behavior: its as-prepared crystals emitted a yellowish green fluorescence but these were transformed into powders which emitted orange-yellow light by grinding. This could be reversed by heating or fuming. It was concluded that the reversible mechanochromism was due to a transformation between crystalline and amorphous states. The other derivative did not exhibit mechanochromism. It was inferred that this derivative had a more planar conjugated structure, leading to a compact intermolecular stacking and strong π-π interactions in the solid state and thus no morphological change during grinding.

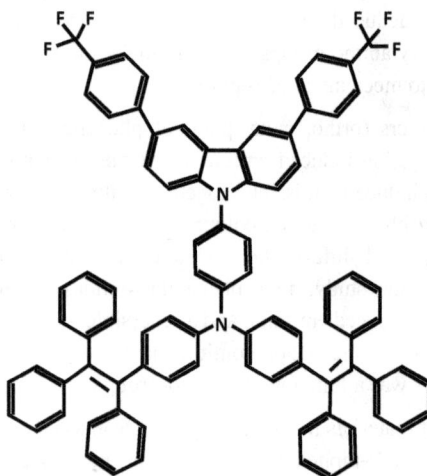

Figure 15. Structure of a carbazole-tetraphenylethene derivative

Three new carbazole-based fluorescent compounds, functionalized with tetraphenylethene (figure 15), have been synthesized[205] which possess a high thermal stability and exhibit various fluorescence behaviours in the solid state. They all also exhibited a marked aggregation-induced emission effect. The molecule shown emitted a blue luminescence which, upon grinding, displayed a new emission peak centred around 510nm and the strong blue emission was replaced by a green luminescence. Upon treating the ground sample with dichloromethane, the green-emitting material reverted to its initial blue color; a cycle which could be repeated again and again. The other two compounds, in which the fluorine grouping was missing, or was replaced by a H_3CO group, exhibited similar mechanochromic behaviours.

A donor-π-acceptor phenothiazine-modified benzothiazole derivative, (E)-3-(4-(benzo[d] thiazol-2-yl)styryl)-10-ethyl-10H phenothiazine, in which phenothiazine groups acted as donors and acceptors (figure 16), was synthesized[206]. The material exhibited a strong yellow fluorescence in solution and in the solid state. As-received crystals exhibited a yellow fluorescence but could be transformed into a powder which emitted orange light by grinding. The fluorescence could be recovered when the ground powder was fumed. This reversible mechanochromic luminescence under grinding and fuming was attributed to a reversible transition between crystalline and amorphous states. Two phenothiazine-based benzoxazole derivatives were similarly synthesized[207] and their properties compared, showing that both compounds emitted a strong fluorescence in the solid state. Ground films of (E)-2-(2-(10-ethyl-phenothiazin-3-yl)vinyl)benzoxazole, without bromine atoms, could self-cure and change back to the original form within 0.25h at room temperature.

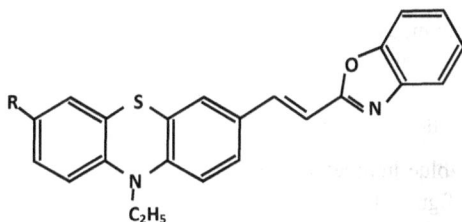

Figure 16. Structure of phenothiazine-based benzoxazole derivatives
R = hydrogen or bromine

The fluorescence of ground films of (E)-2-(2-(7-bromo-10-ethyl-phenothiazin-3-yl)vinyl)benzoxazole exhibited no change for at least 2 weeks. Meanwhile, a higher contrast fluorescence change was observed for a solid film of (E)-2-(2-(7-bromo-10-ethyl-phenothiazin-3-yl)vinyl)benzoxazole, with bromine atoms, under grinding although (E)-2-(2-(10-ethyl-phenothiazin-3-yl)vinyl)benzoxazole and (E)-2-(2-(7-bromo-10-ethyl-phenol thiazin-3-yl)vinyl)benzoxazole exhibited similar red-shifts after grinding. The (E)-2-(2-(10-ethyl-phenothiazin-3-yl)vinyl)benzoxazole and (E)-2-(2-(7-bromo-10-ethyl-pheno thiazin-3-yl)vinyl)benzoxazole have a non-planar phenothiazine component and D–π–A structure. The (E)-2-(2-(10-ethyl-phenothiazin-3-yl)vinyl)benzoxazole is an orange solid that emits an orange emission. A crystal obtained from a mixture of CH_2Cl_2 and n-hexane was yellow and emitted a yellow fluorescence. Grinding produced an orange powder with orange emissions. When the ground powder was exposed to CH_2Cl_2 vapor for some seconds, the orange powder changed rapidly to a yellow one with yellow fluorescence. This fluorescence colour-change could be frequently repeated. Crystals grown from a mixture of CH_2Cl_2 and n-hexane had a maximum emission peak at 537nm, with a shoulder-peak at 560nm. The fluorescence colour was yellow to the unaided eye because of the shoulder-peak. Grinding resulted in an emissive wavelength-change from 537 to 567nm; producing the orange fluorescence of ground powder. When the latter was exposed to CH_2Cl_2 vapour, the orange solid changed rapidly to a yellow one which had the same emission spectrum as that of the crystal. Moreover, the yellow solid could change to the orange form again under mechanical treatment; a cycle which again could be repeated many times. Crystals of (E)-2-(2-(10-ethyl-phenothiazin-3-yl)vinyl)benzoxazole had a wide absorption range in the visible region, with a maximum at 475nm. Other data suggested that (E)-2-(2-(10-ethyl-phenothiazin-3-yl)vinyl)benzoxazole molecules stacked together to form a head-to-tail J-aggregate. The orange solid which resulted from grinding had a blue-shifted (53nm) absorption band, with a maximum at 422nm, as compared with that of yellow crystal; thus suggesting that grinding could change the packing mode of (E)-2-(2-(10-ethyl-phenothiazin-3-yl)vinyl)benzoxazole. By fuming, the absorption spectrum of the orange form could be restored to that of the yellow form. This conversion was repeatable.

Solid-state ultra-violet/blue luminescent materials have been assembled[208] by combining 2,5-diphenyloxazole (figure 17) with four co-assembled building blocks. The two-component molecular solids exhibited changes in fluorescence properties when crystals were ground into powder form. In the case of 2,5-diphenyloxazole-1,4-diiodotetrafluorobenzene, the photoemission band moved from 393 to 384nm, and a shoulder peak appeared at 374nm. Similarly, in 2,5-diphenyloxazole-pentafluorophenol a new shoulder peak occurred at 365nm. But 2,5-diphenyloxazole-4-bromotetrafluorobenzene acquired an entirely different emission spectrum, with a dominant band at 449nm and a shoulder peak at 373nm following grinding. There was no obvious change in the emission spectra for 2,5-diphenyloxazole-octafluoronaphthalene or as-received 2,5-diphenyloxazole upon applying pressure. It was noted that the (020) and (002) diffraction peaks of 2,5-diphenyloxazole-1,4-diiodotetrafluorobenzene and 2,5-diphenyloxazole-pentafluorophenol underwent a shift towards high angles, indicating a decrease in the basal spacing in the (0*l*0) and (00*l*) directions during compression. Such a contraction in lattice separation could produce strong π-π interactions and lead to an increased aggregation of the 2,5-diphenyloxazole chromophores; thus influencing the luminescent properties. In the case of 2,5-diphenyloxazole-1,4-diiodotetrafluorobenzene, the absence of diffraction peaks such as (011) and (111) implied the appearance of amorphous phases during grinding. In the case of 2,5-diphenyloxazole-4-bromotetrafluorobenzene, as well as a shift of the (002) peak to higher degrees, a new peak at 12.4° appeared during grinding thus indicating the formation of a new polymorphic phase of 2,5-diphenyloxazole-4-bromotetrafluorobenzene. This was consistent with the marked shift in the emissions to longer wavelengths. There was no obvious change in the diffraction pattern of 2,5-diphenyloxazole-octafluoronaphthalene following grinding, apart from a decrease in the peak intensities. When ground 2,5-diphenyloxazole-1,4-diiodotetrafluorobenzene   and   2,5-diphenyloxazole-pentafluorophenol powders were exposed to chloroform, their fluorescence emission bands partially returned to the original colours while the associated spectral changes were reversed. When so treating ground 2,5-diphenyloxazole-4-bromotetrafluorobenzene carboxylic acid samples, the emission exhibited no appreciable change apart from the disappearance of the shoulder-band at 373nm. The fluorescence almost completely recovered after recrystallization from methanol. The reversibility of mechanochromic fluorescence was attributed to recoverable intermolecular interactions related to the relative orientations and distances between 2,5-diphenyloxazole and the co-formers.

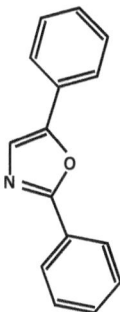

Figure 17. Structure of 2,5-diphenyloxazole

Four triphenylamine or carbazole-based benzothiadiazole molecules have recently[209] been synthesized, among which some donor-acceptor type luminogens exhibited various solid-state fluorescence properties and a reversible high-contrast mechanochromism.

Thiophenes

Multi-functional dithieno[3,2-b;2′,3′-d]thiophene trimers exhibit[210] a marked mechanochromism in lipid bilayer membranes. It is concluded that red-shifts in excitation rather than emission, and fluorescence recovery with increasing membrane order, are consistent with a planarization of the twisted extra-long mechanophores in the ground state.

Luminescent liquid-crystal derivatives containing oligothiophene exhibit[211] a shear-induced phase transition, accompanied by changes in the luminescence colours at room temperature. The shear-induced phases and their colours spontaneously revert to their initial state during aging at room temperature. Ter-thiophene and quarter-thiophene derivatives recovered their original colour faster than did bi-thiophene derivatives. In similar work[212], X-shaped columnar liquid crystals with pyrene as the central core were four-fold conjugated to bi-thiophene and tethered using eight or twelve alkoxy chains at their extremities. The product molecules had hexagonal, tetragonal or rectangular columnar liquid-crystalline structures over a wide temperature range. A mechanochromic photoluminescence change was observed for the columnar form of an X-shaped molecule with 8 alkoxy chains.

Samples of a 5,5′-bis(2-phenylethynyl)-2,2′-bithiophene derivative, containing dendritic constituents linked by amide groups, have been synthesized[213] and had a rectangular columnar structure upon slow cooling from isotropic liquid. On the other hand, a cubic metastable liquid-crystalline phase formed during rapid cooling of the isotropic melt. Mechanical shearing provoked a cubic-columnar phase transition, accompanied by a change in photoluminescence from yellow-green to green. The infra-red data suggested that the shearing led to partial dissociation of the hydrogen bonds, perhaps leading to further disordering of the π-stacks of chromophores.

Mechanochromic luminescent tetrathiazolylthiophene has been investigated[214] as a function of pressure and temperature. Continuous Raman shifts and changes in intensity of the strongest lines of the spectrum occurred upon increasing the pressure or decreasing the temperature, and these spectral changes were attributed to changes in the intermolecular C-H···N hydrogen-bonding. The relevant Raman bands were related to collective normal modes that were associated with C-C bond-stretching of the rings. The response of tetrathiazolylthiophene to various stresses was investigated[215] in such a way that multichromism was obtained over the entire visible region by using a single fluorophore. The differing mechanisms involved in the blue-shift caused by grinding crystals and in the red-shift caused by hydrostatic pressures were investigated. It was concluded that anisotropic and isotropic forms of mechanical loading suppressed and enhanced excimer formation, respectively, in the three-dimensional hydrogen-bonded network.

Polyurethane

Polyurethanes were among the earliest materials to be explored for mechanochromic purposes. The preparation of such polymers generally involves one of two types of process. One is the physical dispersion of some chromophoric additive in the form of a supramolecular aggregate in a pre-formed polymer matrix. The other method involves the covalent insertion of chromophoric units into the macromolecule backbone or side chains. In an early study, structurally and morphologically different segmented polyurethanes which contained a small fraction of conjugated polydiacetylene chains were subjected[216] to tensile straining. The polydiacetylene always oriented themselves into the straining direction during elongation. This implied that the hard segments oriented themselves transversely to the straining direction. The elastomers exhibited mechanochromism in that, during extension, the colour changed from blue to red or yellow. These changes were attributed to a mechanically-induced disordering of the hard domains or to a stress-induced phase transition within those hard domains. In the latter case, reversibility of the phase transition was affected by the soft-segment molecular weight. Entirely reversible phase transitions occurred in segmented polyurethanes which contained crystalline hard domains. Appreciable irreversible hard-domain disruption was detected only at strain levels of more than about 250%.

A mechanochromic elastomer was synthesized[217] by introducing azobenzene chromophores into a co-polyamide oligomer so as to create conformational latches which stabilized the cis-azobenzene components by forming intramolecular hydrogen bonds. The chromophores, introduced into polyurethane, were transformed into the cis-form by ultra-violet irradiation (365nm). The material was then deformed to strains of 100, 200 and 300%. Following 100% strain and relaxation, the absorbance of the co-polymer at 380nm - which revealed transformation to the trans form - had not greatly increased. The absorbance increased at 200 and 300% tensile strain followed by relaxation, demonstrating the existence of a mechanochromic relationship between the absorbance-increase at 380nm in the relaxed state, and the maximum strain to which the sample had been subjected.

Cyano-substituted oligo(p-phenylene vinylene) molecules have been incorporated into thermoplastic polyurethanes by melt processing, to the extent of 0.05 to 0.4wt%, as *in situ* deformation sensors[218]. The emission characteristics were greatly affected by the chemical structure of the chromophore. Only a slight photoluminescence colour-change was observed upon deforming the materials. When 1,4-bis(α-cyano-4-(12-hydroxydodecyloxy)styryl)-2,5- dimethoxybenzene was covalently incorporated into thermoplastic polyurethane, the relaxed material exhibited mainly excimer emission plus an appreciable photoluminescence colour-change during deformation. The mechanochromic response was largely reversible and paralleled the stress-strain response of the material. Segmented polyurethanes were made up of amorphous saturated chain soft segments and rigid π-conjugated hard domains. Within the aggregates of hard domains, π-π interactions could occur and result in perturbation of the opto-electronic properties. Such disruption and restoration of the electronic interactions could lead to an observable mechanochromic response. Oligothiophene diols and di-amines, and a naphthalene di-imide diol were again incorporated into the hard domains of segmented polyurethanes and polyureas while using long poly(tetramethylene oxide) chains as soft segments[219]. The length of the soft segment chains in the segmented polyurethane hindered the electronic coupling of hard domains. A mechanochromic elastomer was again produced[220] by doping bis(benzoxazolyl)stibene into thermoplastic polyurethane to the extent of 0.5%. When film samples were elongated up to 100%, the emission peaks at 475 and 413nm changed in intensity-ratio from 6.3 to 1.8. Upon relaxation, the film length and the emission peaks recovered overall. Short annealing at 120C ensured complete recovery. A further independent study[221] of this material explored the mechanochromic behavior by simultaneously monitoring the applied force, mechanical deformation and fluorescence emissions. A non-linear constitutive model accounted for stress-induced softening. A correlation between the fluorescence response and mechanical deformation demonstrated the feasibility of accurate strain-sensing. It was further confirmed[222] that the optical response was largely reversible and was affected by the initial additive concentration in the polymer.

Shape-memory polyurethane containing covalently-connected 0.1wt% of tetraphenylethylene units exhibited almost 100% shape-recovery and a reversible mechanochromism[223]. Segmented polyurethane elastomer films with diarylbibenzofuranone-based dynamic covalent mechanophores were prepared[224] in which the latter were incorporated into the soft segments. Their cleavage during uniaxial elongation, and their recovery following removal of the stress, were monitored via *in situ* electron paramagnetic resonance measurements and were correlated with marked colour-changes.

Elastomeric polymer film based upon polyurethane, with embedded Rh-OH, was prepared by polycondensation[225]. The rhodamine-based molecule, Rh-OH, exhibited reversible mechanochromic luminescence, but was passive with respect to ultra-violet light. The film exhibited mechanochromism, with a reversible colour change which arose from a change in the Rh-OH molecule from a twisted spirolactam in the closed-ring form to a planarized zwitter-ionic structure in the open-ring state.

A mechanophore comprising a fluorophore-carrying macrocycle and a dumbbell-shaped molecule has been integrated[226] into polyurethane elastomer so that deformation of the polymer produced fluorescence due to spatial separation of the fluorophore and the dumb-bell molecule. This process was immediately reversible and yielded an easily detectable optical signal that was related to the applied force.

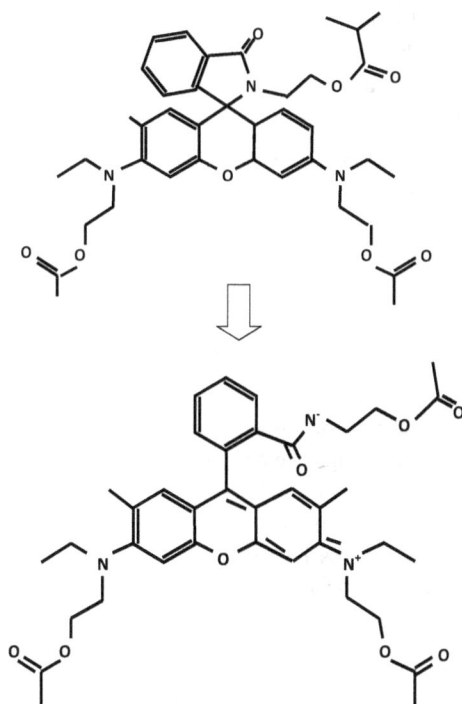

Figure 18. Force-induced ring-opening of a rhodamine-based mechanophore
Upper form: non-fluorescent colourless, lower form: fluorescent coloured

Hydrogels

Polymer gels possess structures which can adapt markedly and rapidly to changes in the local environment, including mechanical force. They are thus eminently mechanochromic. Mechanochromic photonic gels are periodically structured, with a photonic stop-band that can be adjusted by mechanical forces so as to reflect specific colours. One class of mechanosensitive supramolecular interaction involves metal-ligand complexes and metallo-supramolecular polymer gels. Europium can act as a reversible cross-link for telechelic polymers and as a photoluminescent sensor. The diffusion of europium ions through the hydrogel and a spatially heterogeneous formation of cross-links were monitored over time and related to the shape change. Use of a fluoride solution sequestered the europium and removed the cross-links. Prediction of the bend curvature as a function of the number of cross-links formed (according to photoluminescent data) agreed well with the experimental data.

A rhodamine mechanophore-based micellar hydrogel has been found[227] to exhibit excellent mechanochromic and mechanofluorescent properties. Due to the activation of rhodamine spirolactam in the presence of water, combined with a stress concentration effect, the mechanical sensitivity of the material was greatly increased. The stress required to produce the mechanochromism of rhodamine in the hydrogel was then much lower than in its native polymer matrix, and the hydrogel could emit very bright fluorescence under sufficient stress or strain due to a reversible structural change (figure 18).

Reversible mechanochromism has been observed[228] in pyrene-functionalized nanocomposite hydrogels. Highly extendable materials can be prepared by free radical polymerization of poly(N-isopropylacrylamide) and pyrene-terminated polyethylene glycol methacrylate co-polymers; with nanoclay acting as a physical cross-link. When the mechanochromic behavior was investigated using membranes, differential pressures of the order of 100Pa could be detected.

Mechanochromic photonic materials are periodically structured soft materials possessing a photonic stop-band that can be tuned by mechanical forces so as to reflect specific colours. One example[229] of this was a mechano-actuated soft photonic hydrogel which had an ultra-fast response time, a full-colour tunable range, a high spatial resolution and could be actuated by a very small compressive stress. Magnetic assembly, combined with fast photo-polymerization, has been applied[230] to the preparation of mechano-responsive photonic hydrogels. The as-prepared material exhibited colour-switching at a rate of about 1.2nm/ms, across a wavelength range of some 250nm. This was attributed to rapid variation, in response to compressive forces, of the interparticle-separation of one-dimensional Fe_3O_4 nanoparticle chains fixed inside the hydrogel matrix. The chromatic sensitivity to pressure was 20 to 40nm/kPa.

A hydrogel film was prepared[231] from acrylamide and the cross-linker, N,N'-methylenebis (acrylamide), so that the resultant one-dimensional nano-particle chain structure was fixed within the hydrogel under a magnetic field via *in situ* photo-polymerization. These photonic hydrogels had bright colours which could be varied by pressing. The application of 0.2kPa of compression resulted in a 37nm blue-shift of a reflection peak, and the original state could be recovered in less than one second. The sensitivity of the mechanochromic photonic hydrogels could be varied by adjusting the monomer concentration, increasing the detection range in compression from between 0 and 4.3kPa to between 0 and 130.6kPa.

A high-performance mechanochromic photonic gel was developed[232] which was based upon magnetically assembled carbon-encapsulated Fe_3O_4 nanoparticles embedded in a co-polymer of N-hydroxymethyl acrylamide and N-vinylcaprolactam. The carbon-encapsulated Fe_3O_4 content could be as low as 0.18wt%. The material underwent reversible colour changes from purple to red over a wavelength-range of 233nm, with a mechanochromic sensitivity of 53.1nm/kPa and a spatial resolution of less than 100μm.

Materials which can change their reflected colour, depending upon the magnitude of an applied force, could be made[233] from inverse opals by infiltrating colloidal silica particles with a suitable gel. The mechanical sensing range was 17.6 to 20.4MPa. The ratio of the shift in stop-band wavelength, to the change in applied strain, was up to 5.7nm/%. Opals result from the crystallization of monodisperse silica, or are polymer beads of sub-microscopic size. They self-organize into a face-centered cubic lattice from which light is selectively reflected depending upon the wavelength and lattice spacing. This spacing can be changed by deformation, thus constituting an effective mechanochromic effect, as above.

A pressure cell was configured[234] so as to permit dynamic tuning of the transmission characteristics of a mechanochromic colloid-based photonic crystal. The rejection-wavelength underwent a 212nm shift from 645 to 433nm under a pressure of 20kPa. At pressures greater than 5kPa, the rejection-wavelength shifted by about -8nm for every 1kPa increase in pressure.

In the latest proposed strategy for the production of mechanochromic hydrogels, a double-layer design is used[235] in which the two layers contain differing luminescent components: carbon dots and lanthanide ions, having overlapping excitation spectra but distinct emission spectra. Mechanochromism is then exhibited because of a strain-dependent transmittance of the top layer. This controls light emission from the bottom layer and governs the overall hydrogel luminescence. An analytical model can predict the initial luminescence color and color-changes as a function of uniaxial strain.

Nitriles

D-π-A type phenothiazine-modified triphenylacrylonitrile derivatives were synthesized[236] which exhibited aggregation-induced emission. The emission enhancement of derivatives in which the 3-position of phenothiazine was functionalized, was more significant than when the 10-position of phenothiazine was modified. Single-crystal studies revealed that multiple intermolecular interactions, including π-π interactions and hydrogen bonds (C-H···π, C-H···N) locked the molecular conformations, leading to enhanced emission. As-synthesized crystals which emitted yellow, yellowish-orange or yellowish-green light under ultra-violet irradiation were converted into powders which emitted red, orange-red or orange light, respectively, following grinding. These changes were reversed by heating or fuming. The mechanochromism was attributed to a transition between crystalline and amorphous states. The recovery time of ground powders of one derivative was much shorter than those of the others at a given temperature. The latter sometimes did not recover at all at room temperature. The behaviour of the former was attributed to its more twisted conformation, which permitted easier molecular rearrangement of the amorphous state into the crystalline one.

A donor-acceptor fluorophore comprising twisted diphenylacrylonitrile and triphenylamine was synthesized[237] which possessed intramolecular charge-transfer properties. Upon applying hydrostatic pressure, or grinding, the fluorophore exhibited multicolour fluorescence switching, with a clear colour-change from green to red and a change in photoluminescence wavelength of close to 111nm. The large photoluminescence wavelength shift was attributed to a gradual transition of excited states from a local excited state to the charge-transfer state.

A donor-acceptor conjugate, comprising arylamine and two triphenylacrylonitrile units in sterically crowded and highly twisted configurations, was synthesized[238] which exhibited intramolecular charge transfer and aggregation-induced emission. It emitted more than three colours in the solid state during grinding.

Two dicyanodistyrylbenzene molecules were synthesized[239] which exhibited a greatly increased fluorescence emission, and mechanochromic fluorescence changes (tables 4 and 5). In the shorter-wavelength emitting phase, head-to-tail coupling of the local dipoles and multiple C–H...π and C–H...N interactions were held to be responsible for a molecular stacking with rather weak excited-state dimeric coupling. It was found that transformation between two phases occurred during thermal annealing or grinding. The longer-wavelength emitting phase involved a more efficient excited-state dimeric coupling which was attributed to substantial π-π overlap; including anti-parallel coupling of the local dipoles. Upon grinding the blue-emitting phase powder, the fluorescence colour changed immediately to green (509nm). It was noted that the emission spectra of all forms (melt-solidified powder, annealed powder, ground powder) of the other phase exactly matched that of sub-micron sized crystal suspensions.

Table 4. Mechanochromism of
(2Z,2'Z)-2,2'-(1,4-phenylene)bis(3-phenylacrylonitrile

Condition	λ(nm)
as-received powder	441
annealed powder	508
melt-solidified powder	510
ground powder	509
suspension	509

Table 5. Mechanochromism of
(2Z,2'Z)-2,2'-(1,4-phenylene)bis(3-(naphthalen-2-yl)acrylonitrile

Condition	λ(nm)
as-received powder	554
melt-solidified powder	508
ground powder	531
suspension	513

The charge-transfer material, ((Z)-2-(3,5-difluorophenyl)-3-(4-(diphenylamino)phenyl)-acrylonitrile, comprises[240] a diphenylamine donor plus fluorine and cyanogen groups as acceptors and exhibits a multistage mechanochromic luminescence. Orange- and yellow-emitting crystals exhibit a so-called back-and-forth fluorescence response to mechanical solicitation. Modest crushing of the orange or yellow crystals causes a hypsochromic change to cyan (498 to 501nm). Further grinding causes a bathochromic change to green

(540 to 550nm). This was attributed to the molecules being packed by weak interactions, such as C-H$\cdots\pi$, C-H$\cdots$N and C-H$\cdots$F, which facilitated intermolecular charge-transfer in the crystal.

A donor-π-acceptor structured fluorescent material, (Z)-2-(4'-(diphenylamino)-[1,1'-biphenyl]-4-yl)-3-(pyridin-2-yl)acrylonitrile, was synthesized[241] in which cyanogen groups and pyridine were the acceptors and triphenylamine was the donor. The material exhibited a reversible mechanochromic fluorescence change between 552nm and 642nm during pressure-cycling.

Two triphenylacrylonitrile derivatives, having highly twisted conformations, were prepared[242] which exhibited a mechanochromic fluorescent behaviour with red-shifts from sky-blue to light-green occurring during grinding. The ground powders recovered their original state during heating (100C, 60s) or fuming with solvent vapour. These properties were attributed to phase transitions between crystalline and amorphous states.

The biplanar molecule, 2-amino-3-((E)-((2-hydroxynaphthalen-1-yl) methylene) amino) maleonitrile, which consisted of an electron-donor plane and an acceptor plane, was synthesized[243]. It exhibited aggregation-enhanced emission and intramolecular charge transfer, and was weakly emitting when in crystalline form. It exhibited marked emission-enhancement, with a large (67nm) bathochromic shift upon applying mechanical force. Mechanochromic fluorescence arose from crystal defects, where the molecules twisted their configurations so as to break donor-acceptor coupling and lead to strong fluorescence. The defect-induced emission imparted an ultra-high pressure sensitivity to the material, with detection limits as low as 0.62MPa.

Two diaminomaleonitrile-based Schiff bases having a donor-acceptor structure, which exhibited aggregation-enhanced emission, were studied[244]. Emission changes in dimethylamino and dibutylamino versions were determined following the grinding of powder samples. The dibutylamino compound exhibited mechanochromic behaviour, whereas the dimethylamino variant did not. The red powder of the dibutylamino form was converted to yellow by gentle grinding. Its emission color was also changed from red to yellow under ultra-violet irradiation. This change occurred only in ground portions. The addition of THF drops to ground samples partially restored the colour to its original state. Photoluminescence studies showed that the relative peak intensity, rather than the location, was changed by external forces. Single-crystal data suggested that the bulky dibutylamino group led to a loose packing and offered space within which the molecules could adjust their configurations via intramolecular rotation under external constraint. When twisted configurations formed, the π-π interactions were damaged to some extent. The resultant decomposition of J-aggregates and the generation of isolated monomers led to an increase in the intensity of the emission peak at 551nm and to an emission drop at 620nm. The use of THF provoked the recrystallization of the ground sample and the re-formation of J-aggregates. The photoluminescence spectrum was also partially restored to the as-received state. The dimethylamino compound had a more compact stacking and stronger intermolecular π-π interactions than did the dibutylamino compound. This made it difficult to adjust molecular packings, even under external constraint. Further analysis suggested that the as-received dibutylamino form was well-ordered. During grinding, the material was amorphized. Thus the mechanochromic behaviour of the dibutylamino form could be attributed to changes in molecular packing and morphology.

A luminogen, {2-bis[4-(carbazol-9-yl)phenyl]methylene}malononitrile, was synthesized[245] by combining twisted conformations of donor and acceptor units which formed three types of crystal that emitted green (506nm), yellow-green (537nm) and orange (585nm) light under 365nm ultra-violet illumination. The overall emission could be switched between the green, yellow-green and orange colours via morphology modulation by mechanical treatment.

A carbazole derivative, (E)-2-(1-(9-butyl-9H-carbazol-3-yl)-3-(4-fluoro-phenyl)allylidene) malononitrile, containing multiple cyanogen groups and a 4-fluoro-benzene unit, was synthesized[246]. The compound exhibited aggregation-induced emission, due to its highly twisted conformation, as well as solid-state yellow and orange fluorescences. Differing conformations and molecular packings resulted in various mechanochromic behaviors. The fluorescence of the yellow solid underwent a distinct red-shift after grinding, and this was attributed to a break-up of the crystal structure.

Triphenylamine

The propeller-shaped molecule, triphenylamine, has been used[247] to create fluorescent organic materials and relate the effect of structural changes to molecular packing and mechanochromic fluorescence. Thus the substituent, OCH_3, positioned in the triphenylamine phenyl ring, and acceptors such as malononitrile, cyanoacetamide, cyanoacetic acid, ethyl cyanoacetate and diethylmalonate, strongly affect the solid-state and mechanochromic properties as well as the molecular packing. Such studies have revealed that triphenylamine derivatives which lack the OCH_3 substituent exhibit a strong fluorescence and a greater dihedral angle between donor (aminophenyl) and acceptor; leading to weakly- or non-fluorescent products. Substitution at the ortho position increased the dihedral angle. An increase in alkyl groups led to self-reversible high-contrast off-on fluorescence. X-ray diffraction data indicated that induced reversible transformations from crystalline to amorphous were responsible for the fluorescence switching.

Two triphenylamine-based mechanochromic materials, ester-triphenylamine and COOH-triphenylamine, were synthesized[248] by using triphenylamine as the core and benzoic acid and methyl benzoate as the arms. The COOH- version exhibited an obvious red-shift and a transformation from crystalline to amorphous which indicated that the molecular packing had been disrupted.

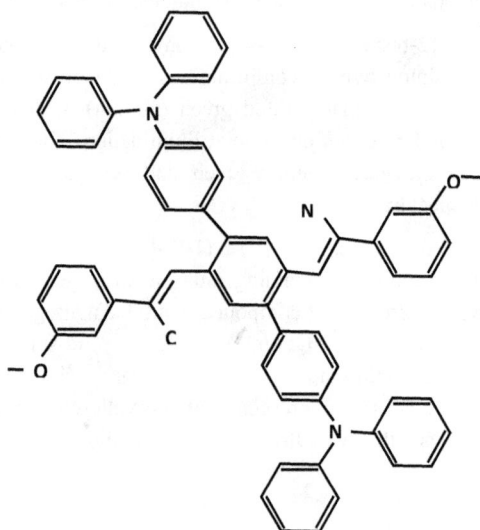

Figure 19. Cruciform structure of DMCS-TPA

Three non-planar D-π-A triphenylamine-based isomers were synthesized[249] which exhibited intramolecular charge-transfer due to electron donation from the triphenylamine to formyl components. The derivatives manifested both reversible mechanochromism and structure-dependent emission. The fluorescence colour could change reversibly between bright blue (455nm), blue (445nm), cyan (479nm) and bright blue-green (497nm), bright blue (460nm), yellowish-green (511nm) due to grinding and dichloromethane-fuming cycles. Single-crystal and powder X-ray diffraction studies suggested that a reversible phase change from crystalline to amorphous was responsible for the observations.

Two isomers comprising twisted triphenylamine and cyanostilbene were synthetized[250], with the fluorescent colour in the crystalline state being sky-blue for one isomer and green for the other. They had very different molecular configurations and packing modes, with the former isomer being planar and the latter being twisted. A reversible mechanochromic fluorescence change was observed in crystalline powders of the planar isomer, with the sky-blue crystals being changed into green-emitting solids by grinding. The original state was recovered by heating (60C, 120s), and the changes were attributed to crystalline-to-amorphization transitions.

Table 6. Mechanochromism of DMCS-TPA Compositions

Composition	Condition	λ(nm)
ortho-DMCS-TPA	as-received	541
ortho-DMCS-TPA	ground	570
meta-DMCS-TPA	as-received	512
meta-DMCS-TPA	ground	599
para-DMCS-TPA	as-received	539
para-DMCS-TPA	ground	590

A donor-acceptor conjugated luminophore, termed DMCS-triphenylamine (figure 19) exhibits[251] aggregation-induced emission and mechanochromic behavior (table 6)[252] with a spectral shift of 87nm. Fluorescence switching from yellowish-green to orange occurs upon compression to just 10MPa or during grinding. The improved mechanochromic results were attributed to extension of the conjugation length and a consequent enhancement of intramolecular charge transfer. The fact that the material is known only as DMCS typifies the common problem faced when attempting to classify or even refer to some of these organic materials.

A dithieno[3,2-b:2',3'-d]pyrrole-based luminogen, which was created[253] by attaching two twisted triphenylamine units, exhibited a reversible mechanochromic luminescence in the solid state and a near-infrared electrochromic switching behavior in solution. A reversible change in fluorescence colour between yellow-green and green was produced by grinding and vapor fuming of the solid powder and was attributed to interconversion between crystalline and amorphous forms.

A fluorescent material was produced[254] by attaching halochromic isoquinoline to a mechanochromic triphenylamine core such as N-phenyl-N-(4-(quinolin-2-yl)phenyl)benzenamine or tris(4-(quinolin-2-yl)phenyl)amine. Both emitted an intense fluorescence in solution and a moderate solid-state fluorescence. Intense grinding of N-phenyl-N-(4-(quinolin-2-yl)phenyl)benzenamine led to a blue-shift in the fluorescence, from 518 to 454nm, while heating reversed the change. The fluorescence of the sample changed from cyan to blue, and back. The fluorescence spectra following slight fracturing of crystals, heavy crushing and heating were studied, showing that heavily crushed and heated samples exhibited equal and higher fluorescence intensities as compared to crystals. Meanwhile, tris(4-(quinolin-2-yl)phenyl)amine did not exhibit any mechanochromism. X-ray diffraction studies suggested that strong grinding transformed N-phenyl-N-(4-(quinolin-2-yl)phenyl)benzenamine from crystalline to amorphous while heating restored it to crystalline form.

Mechanochromic fluorescence has recently[255] been observed in a donor-acceptor structure containing non-planar donor components which permitted tuning of the emission color and control of the emission changes. The fluorescence wavelength could be tuned by changing the strength of the donor-acceptor conjugation, thus yielding a blue to deep-red fluorescence. The twisted conformation, and the continuous internal space which was created by the steric demands of the donor components, permitted rearrangement of the crystal packing during grinding; thus leading to fluorescence changes. Bathochromic and hypsochromic shifts in mechanochromic fluorescence were governed by the presence or absence of stacking interactions with the central acceptor component.

Naphthalimide

The luminescence and colour of naphthalimide could be adjusted[256] by the sonification-induced gelation of an organic liquid using naphthalimide-based agents. Green-emitting suspensions were transformed into orange emitting gels by brief irradiation with ultrasound. When sonification-produced gels were evaporated to give xerogels, the solid-state xerogels exhibited mechanochromism, such that their colour changed from red to yellow and their emissions changed from orange to green upon grinding. This could be reversed by regelation. The mechanochromism of one (figure 20) of the xerogels could be used to gauge mechanical pressures, ranging from 2 to 40MPa, via the fluorescence variations. The results suggested that the various aggregation modes and long-range ordering of the molecules, as regulated by the external stimulus, affected the internal charge transfer of the naphthalimide groups. In this connection, a symmetrical organogelator was designed[257] which contained diacetylene and naphthalimide units. It could self-assemble into a large-scale ordered honeycomb structure, containing micron-sized pores, during evaporation. The micro-pores could be changed into nanopores by adding water via a spontaneous process involving the formation of gel emulsions.

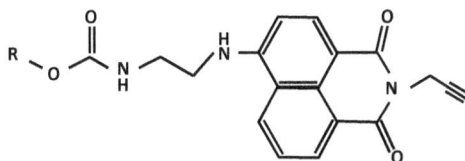

Figure 20. Structure of a pressure-monitoring naphthalimide-based derivative

Mono-alkoxynaphthalene-naphthalimide donor-acceptor materials were studied[258], two of which exhibited a difference in solid-state colour for faster (yellow) and slower (yellow-orange or orange) rates of evaporation from solution. Other materials had only one (yellow-green) colour for both evaporation rates. One of these (figure 21) exhibited a mechanochromic change from orange to yellow in the solid state, with repeatable colour-changing cycling. Structural and spectroscopic data revealed that the behaviour of this material resulted from a 180° molecular rotation in which the thermodynamically more stable head-to-head stacked orange solid reversibly changed into a head-to-tail yellow stacked crystal.

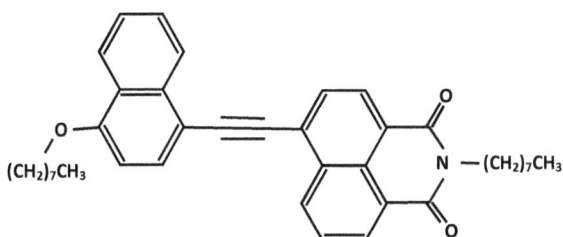

Figure 21. Structure of monoalkoxynaphthalene–naphthalimide compound

When 1,8-naphthalimide, connected to benzoic acid chloride via a methylene group, was separately condensed with 3-amino pyridine and 4-amino pyridine it yielded 4-(1,3-dioxo-2,3-dihydro-1H-phenalen-2-ylmethyl)-N-pyridin-3-yl-benzamide and 4-(1,3-dioxo-2,3-dihydro-1H-phenalen-2-ylmethyl)-N-pyridin-4-yl-benzamide[259]. Their aggregation-enhanced emission behavior depended on the polarity of the solvents. In the as-received form, the 2,3-dihydro-1H-phenalen-2-ylmethyl)-N-pyridin-3-yl-benzamide and 4-(1,3-dioxo-2,3-dihydro-1H-phenalen-2-ylmethyl)-N-pyridin-4-yl-benzamide emitted indigo light at 438 and 428nm, respectively. Upon grinding, the emissions changed to blue with maxima at 469 and 473nm, respectively. These changes were reversed by annealing (220C, 0.25h) the ground samples. The bathochromic shift in emission which was observed was attributed to enhanced conjugation due to transformation of the twisted configuration in the as-received form to a comparatively planar configuration in the ground form.

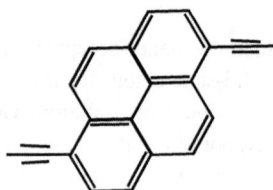

Figure 22. Structure of amphiphilic pyrene luminescent core

Pyrenes

Products with pyrene and rhodamine-B as colour-producing mechanophores, linked by various spacers were synthesized[260], revealing that the one having diphenylalanine as a link underwent a sequential change from deep-blue to bluish-green and thence to reddish. This was associated with a phase-change from gel to xerogel as the solvent evaporated, and to a solid powder, during grinding. The gel and xerogel of the product with a pentaphenylalanine link exhibited a deep-blue colour which changed to bluish-green and then to a reddish powder upon continuously grinding the xerogel. The multi-colour switching was attributed to variations in the structures, which caused a transition of the pyrene excimers from deep-blue to excimer bluish-green, plus a chemical change of rhodamine-B from a spirolactam to a ring-opened red amide. In order to control the colour changes it was deduced to be necessary to constrict the pyrene excimer to an overlapping form of packing.

A water-soluble mechanochromic luminescent pyrene derivative (figures 22 and 23) with two hydrophilic dendrons has been investigated[261]. The two dendritic groups were attached to the luminescent core via amide groups. In order to increase the water-solubility, 32 hydroxyl groups were incorporated into the dendrons at peripheral positions. Grinding of the solid pyrene derivative resulted in a change in the photoluminescence from yellow to green.

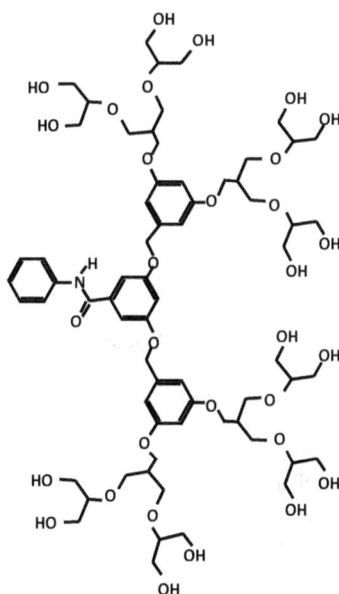

Figure 23. Structure of amphiphilic pyrene hydroxyl group

Exposure to water vapour led to recovery of the original yellow photoluminescence; a cycle which could be repeated at least ten times. Samples of π-conjugated oligomers of high crystallinity were prepared[262] by reacting squaric acid with diaminopyrenes; the oligomers being bonded at the 1,3- and 1,6-positions of the pyrene units: oligo(sq-alt-1,3py) (figure 24) and oligo(sq-alt-1,6py). This markedly affected their planar configuration. Upon grinding the room-temperature solid phases, the colour changed from orange to deep metallic green. This could be reversed by sonification in toluene.

Unexpectedly high-contrast mechanochromic luminescence has been observed[263] upon introducing an ester group into a pyrene skeleton, and materials based upon pyrene-1-carboxylic esters have been investigated (table 7). The photophysical properties of benzyl pyrene-1-carboxylate were such that as-received samples consisted of a white crystalline powder which emitted bright-blue (437nm) fluorescence. Two different types of crystal could be formed by slow evaporation from methanol solvent: colourless needle-like blue-emitting crystals and relatively large light-yellow, green-emitting, crystals. Following grinding, the blue (437nm) emission of the as-received sample changed to green (513nm). Following exposure to hexane vapour, the green emission could be restored to the original colour. This cycling could be repeated several times.

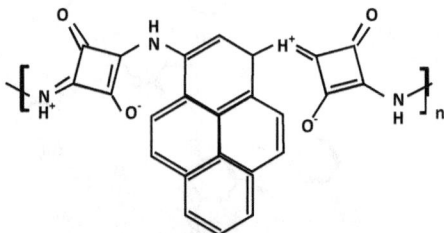

Figure 24. Structure of squaric acid-oligo(sq-alt-1,3py) oligomers

No great difference in mechanochromic behavior was noted upon changing the alkyl group, apart from a minute decrease in the emission wavelength-shift following an increase in alkyl length. The mechanochromic luminescence emission wavelength was red-shifted from 67 to 72nm across the range of esters. It was assumed that the molecular stacking of these esters was controlled mainly by the π-π interaction, rather than by other molecular interactions; thus resulting in the small influence of alkyl length upon the mechanochromic behaviour. On the other hand, methyl pyrene-1-carboxylate was among the smallest mechanochromic fluorophores known to exhibit an obvious mechanochromic behaviour. In the case of the above green-emitting crystals, single-crystal structural analysis revealed that a molecule-pair stacking with a face-to-face orientation was present, with a vertical distance of about 3.52Å between adjacent pyrene rings, a pitch angle of about 64° and a roll angle of about 86°. There was a strong π-π interaction between adjacent pyrene rings in the green-emitting crystals.

Table 7. Mechanochromism of pyrene-1-carboxylic esters

Ester	Condition	λ(nm)
Benzyl pyrene-1-carboxylate	as-received	355
Benzyl pyrene-1-carboxylate	ground	437
Methyl pyrene-1-carboxylate	as-received	352
Methyl pyrene-1-carboxylate	ground	440
Ethyl pyrene-1-carboxylate	as-received	352
Ethyl pyrene-1-carboxylate	ground	441
Butyl pyrene-1-carboxylate	as-received	352
Butyl pyrene-1-carboxylate	ground	445
Octyl pyrene-1-carboxylate	as-received	353
Octyl pyrene-1-carboxylate	ground	441
Pyren-1-yl acetate	as-received	324
Pyren-1-yl acetate	ground	451

Methyl pyrene-1-carboxylate exhibited mechanochromism with an emission wavelength red-shift of 72nm but its isomer, pyren-1-yl acetate, exhibited a mechanochromic behavior which was similar to that of the unmodified pyrene compound, with an emission-wavelength red-shift of only 19nm. This suggested that direct introduction of an ester group into the pyrene ring was essential in order to impart an obvious mechanochromic luminescence. Single crystals of pyren-1-yl acetate were obtained by slow evaporation of hexane solution at room temperature, giving a packing with a face-to-face orientation with a vertical distance of about 3.32Å between adjacent pyrene rings, a pitch angle of about 35° and a roll angle of about 78°; indicating an obvious π-π interaction between adjacent pyrene rings. Because of the relatively strong electron-withdrawing tendency of the ester groups, it had been assumed that those groups could suppress the π-π interaction of pyrene rings and render the molecular stacking relatively loose. This would have produced a blue-shifted emission in the as-received state and a large emission-wavelength shift during grinding. The relatively strong π-π interaction of pyren-1-yl acetate limited slip of the molecular stacking, and only a small emission-wavelength shift occurred.

Bis-pyrene derivatives (figure 25) have been synthesized[264] in which the two pyrene units were connected by para-phthaloyl or meta-phthaloyl links. The solid para-phthaloyl compound did not respond to mechanical shearing. The fluorescence emissions of supramolecular aggregates of the solid meta-phthaloyl compound exhibited a reversible mechanochromic behaviour. The differing molecular symmetries which arose from the different substitution-positions in the para-phthaloyl and meta-phthaloyl compounds produced differing packing densities and corresponding intermolecular interactions, thus resulting in the contrasting responses to mechanochromic stimuli. The intermolecular interactions of the para-phthaloyl form, with its high molecular symmetry, were too strong to respond to such stimuli. On the other hand, the intermolecular interactions of the meta-phthaloyl compound were conducive to mechanochromic fluorescence. It was concluded that the more symmetrical para-phthaloyl form, with its para-substitution, was packed so densely, with strong intermolecular interactions, that it was difficult for it to respond to external stimuli. The meta-phthaloyl form, with its lower molecular symmetry, could permit appropriate intermolecular interactions and maintain its well-organized structure. It was however subject to external forces, thus resulting in the mechanochromic fluorescence.

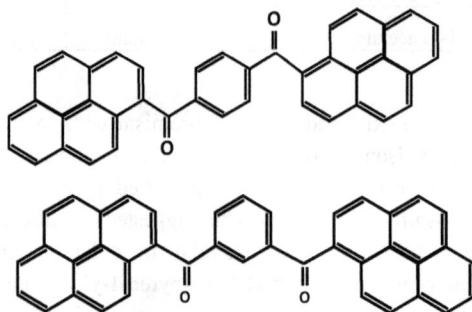

Figure 25. Structure of bis-pyrene derivatives with a (above) para-phthaloyl or (below) meta-phthaloyl link

Miscellaneous Polymers and Composites

A radical-type mechanochromic polymer was developed[265] which exhibited a colour-change from white to green due to the dissociation of a diarylbibenzothiophenonyl unit at the mid-point of a polystyrene chain. Materials which could exhibit various colours under mechanical stressing, such as a green colour due to shear, could then be formulated simply by mixing radical-type mechanochromic polymers which emit primary colours. Combinations with other mechanochromic polymers exhibited different colours under similar shear forces, and could then offer a so-called rainbow mechanochromism.

Diacetylene-containing segmented copoly(ether urea) was prepared[266] so as to include a reactive diacetylene spacer within the hard domains of the segmented elastomer. It was cross-polymerized topochemically by means of ultra-violet radiation, without affecting the flexibility of the material, leading to a bright-blue insoluble product. The cross-polymerized material could be caused to change colour by mechanical means. Study of the mechanochromic colour-change from blue to yellow during elongation showed that the change was irreversible at strains greater than 80%. During deformation, the polydiacetylene chains became oriented parallel to the deformation axis.

The introduction of a liquid-crystalline photo-emissive compound into a polymethacrylate with photo-emissive side-groups imparts[267] repeatable change possibilities with regard to the photo-emission wavelength following grinding and annealing. Colour-changes of a handwriting sample demonstrated the introduction of an anisotropic photo-emission behavior, parallel to the grinding direction, into a composite polymeric film.

Polymer-inorganic composites comprising diarylbibenzofuranone components have been studied[268]. The central C-C bonds of diarylbibenzofuranone can be cleaved by mechanical force to produce stable blue radicals. Polymer-inorganic composite materials having a rigid networks, and prepared from diarylbibenzofuranone alkoxysilane derivatives, exhibited a significant activation of the incorporated diarylbibenzofuranone linkages during grinding, with sensitivities which were more than 50 times higher than that those of diarylbibenzofuranone monomers. The increased sensitivity was attributed to the effective transmission of mechanical force to the diarylbibenzofuranone units of the network and to the suppressed recombination of by the rigid framework. When rigid frameworks were incorporated into elastomers, the diarylbibenzofuranone mechanophores at the interface between organic and inorganic domains were preferentially stimulated by elongation.

The synthesis of mechanically stable and self-healing superlattice nanocomposites is possible[269] via the self-assembly of single-component so-called sticky polymer-grafted nanoparticles. Hydrogen-bonding interactions between the nanoparticles provided sufficient cohesive energy to bind nanoparticles into strong materials. In addition, mechanochromic behaviour was manifested as a blue-shift during sample-stretching. It was a result of the contraction of the spacing between particle planes. The wavelength of reflected light changed from 625 to 550nm upon straining a sample by 30%. The maximum wavelength that could be attained was 550nm and did not change further. Before reaching a plateau, the wavelength of the reflected light was changing by 2.4nm per percent of strain.

It has been shown[270] that an epoxy vitrimer, containing 4-aminophenyl disulfide, exhibited mechanochromic properties. When hit with a hammer, a green coloration was observed. Simple bending of a composite did not produce any color change. The mechanochromic effect occurred only when resin-breaking was involved. When the plain resin powder was ground, the same green coloration appeared. These effects were attributed to the formation of sulfenyl radicals. The latter disappeared within 24h at room temperature, or within a few seconds when heated to above the glass transition temperature. This was probably due to recombination of the radicals. No mechanochromic effect was observed in a similar epoxy vitrimer system which was made from 2-aminophenyl disulfide. Theory suggested that the difference in behavior between the two systems was due to a greater change in the dipole moment when sulfur and nitrogen atoms were in para positions.

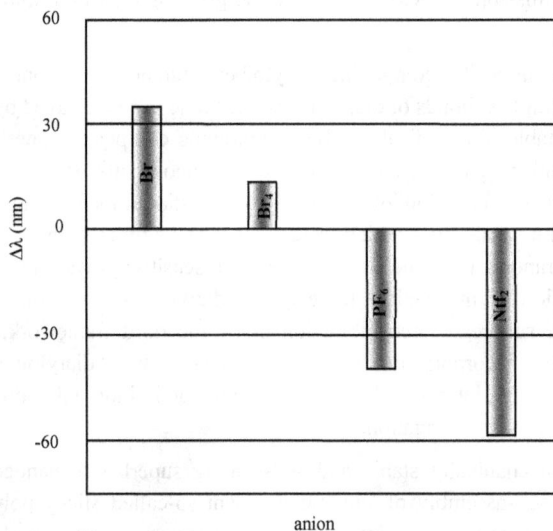

Figure 26. Effect of counteranion type on wavelength change of phosphonium salts during grinding

Diamine-cured epoxies which contain 4,4'-diaminodiphenylmethane undergo changes in both absorption and fluorescence[271] in that samples change from an original blue to a mechanochemically-activated red fluorescence under uniaxial compression right from the early stages of strain-hardening. Orange and green chromophores are also generated by compression: the orange chromophore being a red-emitting fluorophore, while the green chromophore is non-fluorescent. The 4,4'-diaminodiphenylmethane structure is the cause of the mechanochromic response. The red-emitting orange chromophore was suggested to be a reactive radical intermediate of the core 4,4'-diaminodiphenylmethane structure, while the green chromophore was a quinoidal methine resulting from intermediates.

The combination of an epoxy and amine curing-agent, tetraglycidyl-4'-4 diamino diphenylmethane with diethylenetriamine, can exhibit[272] mechanochromic behaviour. The product can produce a strong fluorescence, under compression or impact, which is detectable at about 0.5GPa and visible at about 1.0GPa; increasing monotonically with increasing stress. The fluorescence intensity decays with time, thus suggesting that the emission originates from mechanochemically-created radical species which slowly recombine to form non-fluorescent species.

Phosphonium

By using a single tributyl(perylen-3-ylmethyl)phosphonium fluorophore, marked dual-mode mechanochromic changes can be triggered by adjustable cation-anion interactions[273]. The photophysical properties of as-received samples of tributyl(perylen-3-ylmethyl)phosphonium were first investigated, where the phosphonium salts involved various counter-anions, such as Br^-, BF_4^-, PF_6^- and NTf_2^- (bis((trifluoromethyl)sulfonyl)amide), and all exhibited luminescence in the solid state. Various emissions were observed from as-received samples of these materials and these were in good agreement with the order of strengths of the cation-anion interactions, thus implying that the molecular stacking of tributyl(perylen-3-ylmethyl)phosphonium could be controlled by changing the counter-anion. All of these phosphonium compounds also exhibited good mechanochromic behaviours (table 8). Following grinding, red-shifts of the emissions were observed for some counter-ions while blue-shifts were observed for others (figure 26).

Table 8. Mechanochromism of tributyl(perylen-3-ylmethyl)phosphonium compounds

Derivative	Condition	λ(nm)
tributyl(perylen-3-ylmethyl)phosphonium-Br	as-received	525
tributyl(perylen-3-ylmethyl)phosphonium-Br	ground	560
tributyl(perylen-3-ylmethyl)phosphonium-Br$_4$	as-received	547
tributyl(perylen-3-ylmethyl)phosphonium-Br$_4$	ground	560
tributyl(perylen-3-ylmethyl)phosphonium-Br$_6$	as-received	598
tributyl(perylen-3-ylmethyl)phosphonium-Br$_6$	ground	558
tributyl(perylen-3-ylmethyl)phosphonium-Ntf$_2$	as-received	622
tributyl(perylen-3-ylmethyl)phosphonium-Ntf$_2$	ground	563

The mechanochromic luminescences of the compounds were also in good agreement with the order of strength of the cation-anion interaction. This suggested that the cation–anion interaction could partially govern the mechanochromic behaviour. Paralleling the order of strength of the cation–anion interactions, $Br^- > BF_4^- > PF_6^- > NTf_2$, the emission wavelength changes decreased in the range of red-shifted mechanochromic luminescence for the anions, Br^- and BF_4, while the emission wavelength changes increased in the range of blue-shifted mechanochromic luminescence for the anions, PF_6^- and NTf_2. As-received samples of tributyl(perylen-3-ylmethyl)phosphonium-Br were yellow solids which exhibited a yellow-green (525nm) emission. Following grinding, the emission changed from yellow-green to orange (560nm). Annealing at about 100C for some minutes changed the orange back to yellow-green. The orange colour could be again recovered by re-grinding. The emissions were completely reversible between the two states by repeated grinding and annealing. Upon changing the counter-anion to bis((trifluoromethyl)sulfonyl)amide, the as-received solid material was an orange crystalline powder, with a red emission at 622nm. With regard to the mechanochromic luminescence, a marked blue-shift (59nm) of the emission was observed following grinding. Reversible emission changes between red (622nm) and orange (563nm) occurred during repeated cycles of grinding and annealing. Various observations suggested that oriented molecular stackings existed in as-received samples of each compound. Compared to sample in the as-received state, only a few broad diffraction signals with weak intensities were found for ground samples, suggesting a crystalline to amorphous phase transition during grinding, which could be reversed by heat treatment.

Study of single crystals of tributyl(perylen-3-ylmethyl)phosphonium-Br revealed a stacking of the molecules, in which adjacent perylene planes were arranged in a crossed orientation with an angle of about 31° between the ring normal and the vector between the ring centroids; together with a centroid-centroid distance of 6.91Å. No obvious π-π interactions between the perylene rings were detected. The luminescence decay of as-received and ground samples had lifetimes of 2.46 and 2.57ns, respectively. Aggregation of the excimer could be excluded during mechanochromism, due to the similar and relatively short lifetimes in the differing states. It was concluded that the red-shifted orange emission could be attributed to increased π-π interaction in the amorphous phase, and the original yellow-green emission with weak π-π interaction could be recovered by annealing. The luminescence decay of as-received and ground tributyl(perylen-3-ylmethyl)phosphonium-Br$_4$ exhibited lifetimes of 3.13 and 4.48ns, respectively. A mechanochromic behaviour similar to that of tributyl(perylen-3-ylmethyl)phosphonium-Br was posited, and a red-shifted emission following grinding might be attributed to the enhancement of π-π interactions in the amorphous phase. As-received single crystals of tributyl(perylen-3-ylmethyl)phosphonium-NTf$_2$ were orange, with red emissions. X-ray diffraction data showed that the molecules were stacked in the same way as in tributyl(perylen-3-ylmethyl)phosphonium-Br, but a face-to-face arrangement was observed for the adjacent perylene planes, with a vertical distance of about 3.53Å between adjacent perylene rings, a pitch angle of about 76° and a roll angle of about 75°, with strong π-π interactions between adjacent perylene rings. Compared to tributyl(perylen-3-ylmethyl)phosphonium-Br, as-received and ground samples of tributyl(perylen-3-ylmethyl)phosphonium-NTf$_2$ exhibited relatively long lifetimes of 41.39 and 26.26ns, respectively.

Taken together with strong π-π interactions in as-received samples, it was deduced that an excimer-like molecular aggregate might form in as-received samples of tributyl(perylen-3-ylmethyl)phosphonium-NTf$_2$ leading to a long-wavelength red light emission. During grinding, the relatively well-defined microcrystalline structures were amorphized and the excimer-like molecular aggregation was completely or partially destroyed, leading to a blue-shifted orange emission. The initial crystalline state of tributyl(perylen-3-ylmethyl)phosphonium-NTf$_2$ could be regenerated by annealing, together with the red emission. Single crystals of tributyl(perylen-3-ylmethyl)phosphonium-PF$_6$ were yellow, and the crystals were again stacked as molecular-pairs with a vertical distance of some 3.44Å between adjacent perylene rings. Instead of the face-to-face arrangement of the adjacent perylene planes of tributyl(perylen-3-ylmethyl)phosphonium-NTf$_2$, a crossed orientation was observed for the molecules in crystals of tributyl(perylen-3-ylmethyl)phosphonium-PF$_6$. The crossed orientation of adjacent perylene rings resulted in relatively weak π-π interactions between adjacent perylene rings in crystals of tributyl(perylen-3-ylmethyl)phosphonium-PF_6; in good agreement with the blue-shifted emission of as-received samples.

Table 9. Mechanochromism of tributyl(pyren-1-ylmethyl)phosphonium compounds

Derivative	Condition	λ(nm)
tributyl(pyren-1-ylmethyl)phosphonium-Br	as-received	400
tributyl(pyren-1-ylmethyl)phosphonium-Br	ground	484
tributyl(pyren-1-ylmethyl)phosphonium-Br_4	as-received	398
tributyl(pyren-1-ylmethyl)phosphonium-Br_4	ground	483
tributyl(pyren-1-ylmethyl)phosphonium-Br_6	as-received	399
tributyl(pyren-1-ylmethyl)phosphonium-Br_6	ground	486
tributyl(pyren-1-ylmethyl)phosphonium-Ntf_2	as-received	464
tributyl(pyren-1-ylmethyl)phosphonium-Ntf_2	ground	473

The blue-shifted mechanochromic emission and relatively long lifetimes, of 62.16ns and 15.35ns, of as-received and ground states, respectively, suggested that tributyl(perylen-3-ylmethyl)phosphonium-PF_6 could undergo a mechanochromic mechanism similar to that of tributyl(perylen-3-ylmethyl)phosphonium-NTf_2, with complete or partial destruction of the excimer-like molecular aggregates formed in as-received samples during grinding. X-ray diffraction for tributyl(perylen-3-ylmethyl)phosphonium-Br, tributyl(perylen-3-ylmethyl)phosphonium-PF_6 and tributyl(perylen-3-ylmethyl)phosphonium-NTf_2 also showed that π-π interactions between perylene rings increased upon decreasing the strength of the cation–anion interaction. In the range of relatively strong cation–anion interactions in tributyl(perylen-3-ylmethyl)phosphonium-Br and tributyl(perylen-3-ylmethyl)phosphonium-BF_4, relatively weak π-π interactions between the perylene rings were observed, leading to emissions of relatively short wavelength in as-received samples and a red-shifted mechanochromic emission following grinding. On the other hand, relatively strong π-π interactions led to red-shifted wavelengths in as-received samples and the blue-shifted mechanochromic emissions of tributyl (perylen-3-ylmethyl) phosphonium-PF_6 and tributyl (perylen-3-ylmethyl) phosphonium-NTf_2. A similar study[274], involving materials based upon tributyl(pyren-1-ylmethyl)phosphonium, had yielded analogous results (table 9).

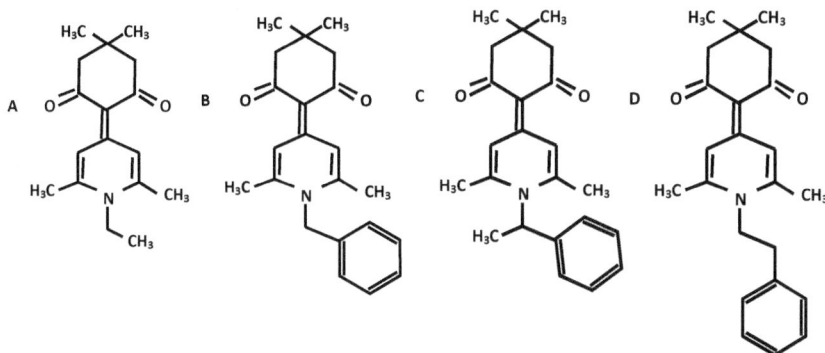

Figure 27. Structures of 1,4-dihydropyridine derivatives

Pyridines

Four 1,4-dihydropyridine derivatives (figure 27) were synthesized[275] which contained various N-substituted groups. One compound, which contained a 2-phenylethyl group, exhibited a reversible high-contrast mechanochromism. Compounds which contained an ethyl group or a 1-phenylethyl group exhibited no mechanochromic behaviour.

A compound which contained a benzyl group took two different crystalline forms, in which a blue-emitting form was mechanochromic, while a green-emitting form was non-mechanochromic (table 10). It was concluded that N-substituted groups were decisive in imparting mechanochromism. Phase transformation between crystalline states was deemed to be responsible for the reversible mechanochromic properties (figure 28) of some of the compounds. Red-shifts of the fluorescence spectra were attributed to planarization of the molecular conformations and to a resultant enhancement of the degree of π-electron conjugation.

The luminescence of 4,4'-bipyridine in crystalline and powder forms was studied[276] under high pressure by using a diamond anvil cell. In single crystals, the molecules were arranged in a parallel fashion, with C–H···N and π···π interactions. In the powder form, the intermolecular interactions were greatly diminished. The crystals exhibited a clear bathochromic shift of the emission band, whereas powder samples maintained a fixed luminescence band during compression. The crystals exhibited a better mechanochromic behaviour than did the powder, because the intermolecular interactions predominated in the crystal.

Table 10. Colour changes of ground and fumed 1,4-dihydropyridine derivatives

Derivative	Condition	λ(nm)
A	as-received	455
A	ground	459
B	as-received	463
B	ground	482
B	fumed	463
B	as-received	490
B	ground	492
C	as-received	486
C	ground	487
D	as-received	450
D	ground	498
D	fumed	448

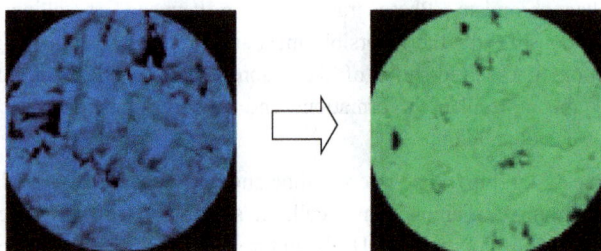

Figure 28. Example of colour-change: derivative D

Materials Research Forum LLC
https://doi.org/10.21741/9781644900277

Several series of N-alkylated 1,4-dihydropyridine derivatives comprising various electron-withdrawing end-groups have been synthesized[277]. The electron-withdrawing groups affected the mechanochromic properties, and polymorphs of the 1,4-dihydropyridine derivatives exhibited a decreasing trend as the length of the alkyl chain increased; thus suggesting that longer alkyl chains were inimical to polymorph formation. Differences in the emissions of the polymorphs were attributed mainly to their differing intermolecular interactions and molecular packings. As-received samples of 2-(1-ethyl-2,6-dimethylpyridin-4(1H)-ylidene) malononitrile emitted a bright-blue (461nm) fluorescence. Two different structures of 2-(1-ethyl-2,6-dimethylpyridin-4(1H)-ylidene) malononitrile could be crystallized. One emitted a cyan (470nm) fluorescence whereas a lawn-green crystal produced a redder (489nm) emission.

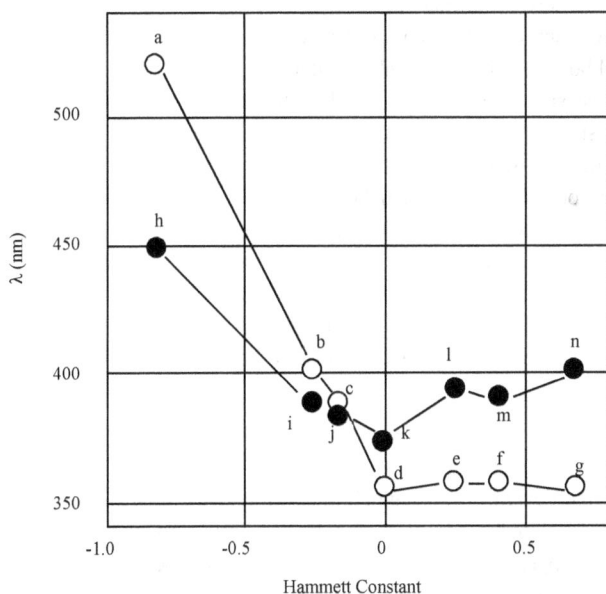

Figure 29. Effect of substituents on the two rings of C2-aryl (open circles) and C7-aryl (closed circles) [1,2,4]triazolo[1,5-a]pyrimidine. a) 4-(CH₃)₂NPh, b) 4-CH₃OPh, c) 4-CH₃Ph, d) Ph, e) 4-ClPh, f) 4-CHOPh, g) 4-CNPh, h) 4-(CH₃)₂NPh, i) 4-CH₃OPh, j) 4-CH₃Ph, k) Ph, l) 4-ClPh, m) 4-CHOPh, n) 4-CNPh

A fluorescence colour-change from blue to yellow-green occurred when the original 2-(1-ethyl-2,6-dimethylpyridin-4(1H)-ylidene) malononitrile was ground. When the ground sample was exposed to ethyl acetate vapor, its fluorescence returned to that of the as-received sample. These results indicated that the mechanochromism could be attributed to a phase transition between various crystalline states, rather than to the familiar transformation between crystalline and amorphous states. Investigation of the response of the above crystals to grinding showed that gentle grinding produced a blue (459nm) emission whereas a yellow-green (502nm) emission resulted from more intense grinding. These results demonstrated that the emissions of 2-(1-ethyl-2,6-dimethylpyridin-4(1H)-ylidene) malononitrile could be switched between four colors via transformations between four crystalline forms.

In the most recent study[278], mechanochromic luminescence was induced in a complex of mechano-inactive compounds, such as dye/acid complexes containing the same π-conjugated backbone. The luminophore underwent blue and red shifts of the photoluminescence spectra when combined with various acids by grinding, but itself exhibited only a slight mechano-response. The preparation of compounds having similar molecular backbones to those of the dye/acid complex helped to clarify the color-change mechanism. The compounds exhibited both blue and red shifts in photoluminescence, and a diffuse reflectance spectra after grinding. This suggested that mechanochromic luminescence in the hydrogen-bonded complex resembled its monomeric analogue, and that the aggregation structure played the important role in the mechanoresponse, rather than the π-conjugated structure. A colour change could mechanically induced by imitating the solid-state aggregate structure of other mechanoresponsive compounds, without synthetic modification.

Pyrimidines

One strategy for the design of mechanochromic luminogens has been based[279] upon the dipole moment of donor-acceptor molecules. A route to synthesizing 2,7-diaryl-[1,2,4]triazolo[1,5-a]pyrimidines involved C-H arylation. The materials, with electron-donating groups on the 2-aryl unit and electron-withdrawing groups on the 7-aryl unit possessed a relatively small dipole moment and red-shifted mechanochromism. When those two aryls were interchanged, the resultant luminogens had a relatively large dipole moment and exhibited a blue-shifted mechanochromism. In one particular case, where 4-$(CH_3)_2NPh$ and 3,5-$(CF_3)_2Ph$ were the interchangeable aryls, grinding of the as-received powder produced a marked red-shift, with an emission color-change from yellowish-green (539nm) to orange-red (588nm) (figure 29). The material with aryls reversed emitted, in the as-received condition, red with a maximum at 620nm. Upon grinding, the

Materials Research Forum LLC

https://doi.org/10.21741/9781644900277

emission blue-shifted to orange-yellow (563nm). This showed that luminogens with interchanged and rings exhibited opposite mechanochromic trends. Differential scanning calorimetry showed that the ground samples exhibited an exothermal peak at about 70 and 76C, respectively, revealing a phase transition between the metastable ground state and the more stable original state. The emission of one of the two materials could be reversibly changed between states by repeated grinding and annealing, with no obvious fatigue, and the emission of ground samples could be restored by solvent fuming. The emission of ground samples of the other material could be restored to 582nm only by annealing.

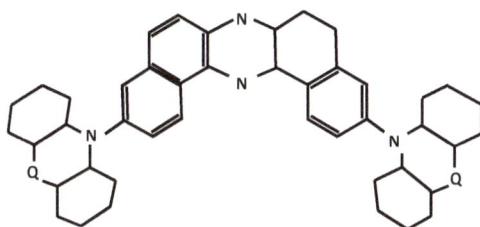

Figure 30. Structure of dibenzo[a,j]phenazine-dihydrophenophosphanizine sulfide

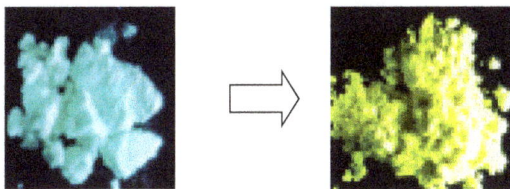

Figure 31. Colour-change caused by grinding

Pyrimidine-based BF_2 complexes which exhibited aggregation-induced emission and mechanochromic luminescence have been developed[280]. They exhibited an intense fluorescence in the aggregated/solid-state which resulted from a large Stokes shift and aggregation-induced emission. Weak intermolecular interactions were held to be responsible for the intense fluorescence in the solid state. In addition to the marked aggregation-induced emission behaviour, there was also a reversible colour-change response to grinding, with a distinct red-shift. The emission band of powdered material having the formula, $C_{30}H_{22}BF_2N_3O$ exhibited a red-shift from 491 to 509nm during grinding; sufficient to produce a visible color change from green to chartreuse. When the pressed powder was exposed to CH_2Cl_2 vapour for 600s, the emission band was blue-shifted to 491nm, and the original green colour was restored. The as-received powder gave intense X-ray diffraction peaks due to its crystal structure. Under pressure, a new set of peaks appeared; indicating that the powder had been converted to another crystal form by the grinding. The mechanochromism was attributed to the propeller-like configuration and to the donor–acceptor nature of the material. Grinding was imagined to lead to changes in packing, leaving a more planar structure and better dipole–dipole intermolecular interactions and a large red-shift of the fluorescence. All of the boron atoms adopted a tetrahedral geometry in order to form N^O-chelate six-membered rings.

Phenazine

Donor-acceptor-donor π-conjugated molecules which comprise dibenzo[a,j]phenazine as an acceptor, and phenothiazines as donors, have been developed[281] which exhibit tricolour mechanochromic luminescence and a thermally-activated delayed fluorescence. It is found that so-called two-conformation switchable phenothiazine units play a decisive role in imparting multi-colour mechanochromic luminescence.

A twisted D–A–D triad material (figure 30) was synthesized[282] from dibenzo[a,j]phenazine as the acceptor and dihydrophenophosphanizine sulfide as the donor. It exhibited mechanochromic luminescence. In the solid state, crystals exhibited a blue–green (497nm) emission upon irradiation with ultra-violet light. Grinding caused a morphological transition from a crystalline state to an amorphous material which emitted yellow light (figure 31). Fuming or annealing led back to a crystalline state.

Cyanostilbene

Five twisted donor-π-acceptor cyanostilbene derivatives which contained carbazole were synthesized[283], showing that their emission could be varied from blue to orange by changing the donor. They exhibited typical intramolecular charge transfer, aggregation-induced emission and crystallization-induced emission enhancement, with hydrogen-, methyl- and chlorine-substituted materials, in particular, manifesting reversible mechanochromism with emission-changes of up to 119nm. On the other hand, methoxyl- and nitryl-substituted materials exhibited no mechanochromism, due to their marked crystallization tendency. The mechanochromism was attributed to the existence of stator-rotor structures, twisted conformations and crystal-packing modes. Substitution thus

Materials Research Forum LLC
https://doi.org/10.21741/9781644900277

offers an effective means for obtaining full-colour materials which exhibit mechanochromic properties.

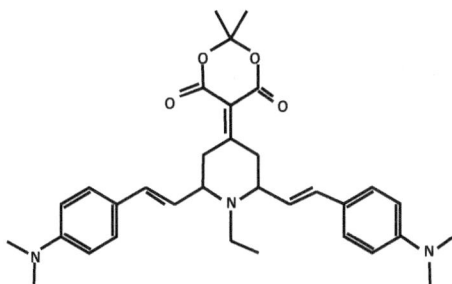

Figure 32. Structure of 5-(2,6-Bis((E)-4-(dimethylamino)styryl)-1-ethylpyridin-4(1H)-ylidene)-2,2-dimethyl-1,3-dioxane-4,6-dione

Table 11. Summary of colour changes of miscellaneous materials

Material	Pre-Grinding	Post-grinding
N-methyl fluorenylidene-acridane	yellow	dark green
vinamidinium with phenyl units	yellow	green-yellow
vinamidinium with p-bromophenyl units	orange	yellow
(E)-3-chloro-4-((4-(diethylamino)-2-hydroxy benzylidene)-amino)benzoic acid	green	orange-red
fluorenyl-containing tetra-substituted ethylene	green	blue
4-[bis(4-methylphenyl)amino]benzaldehyde	light-blue	green-yellow
Triphenylamine-(2,2-dicyanovinyl)phenyl	green	orange-red
(E)-6-((4-(diethylamino)-2-hydroxybenzylidene)amino)-2-naphthoic acid	green	orange-red
boroquinol	light-blue	yellow
N-phenylcarbazol-substituted tetra-arylethene	sky-blue	cyan
tetrakis(4-(dimethylamino)phenyl)ethylene	blue	yellow-green

A highly-fluorescent difluoroboron complex containing α-cyanostilbene and thiophene has been synthesized[284], revealing that a rigid structure, intermolecular hydrogen-bonding, good ordering and a loose molecular packing were the main causes of the high fluorescence emission. The colours and fluorescence emission could be modified by external stimuli such as grinding. The as-synthesized material was a yellow crystalline powder which emitted a bright fluorescence (532nm). Following grinding, an orange powder with weak orange-yellow (542nm) fluorescence was found. Grinding caused a red-shift of about 10nm. When dichoromethane was dropped onto the powder, the color and fluorescence of the ground material recovered immediately. This cycle could be repeated many times. Crystalline-amorphous transformations between as-received and ground states were suggested to explain the mechanochromic luminescence changes.

Diones

Two soluble 9-anthryl-capped pyrrolo[3,4-c]pyrrole-1,4(2H,5H)-dione derivatives, with benzene and thiophene as spacing units, were synthesized[285] to give 1,4-diketo-2,5-dioctyl-3,6-bis(9-anthrylphenyl)pyrrolo[3,4-c]pyrrole and 1,4-diketo-2,5-dioctyl-3,6-bis(9-anthryl thiophen)pyrrolo[3,4-c]pyrrole. In solution, these products emitted strong and weak fluorescence, respectively, but their crystals both emitted strong fluorescence; implying that only the latter exhibited aggregation-enhanced emission although both had a twisted backbone. Grinding and pressing changed the fluorescence of the former crystalline material from yellow to red. The transition between crystalline and amorphous states was suggested to explain the mechanochromic changes.

5-(2,6-bis((E)-4-(dimethylamino)styryl)-1-ethylpyridin-4(1H)-ylidene)-2,2-dimethyl-1,3-dioxane-4,6-dione (figure 32) exhibits[286] aggregation-induced emission due to its highly twisted configuration. It has three crystalline polymorphs, which emit yellow, orange and red fluorescence, and these differing emissions depend mainly upon their molecular configuration due, in turn, to weak intermolecular interactions. The crystalline polymorphs also exhibit differing mechanochromic behaviours under pressure. For instance, when the orange polymorph was gently ground, it changed to bright yellow; a blue-shift of 598 to 570nm. Upon increasing the pressure on the same sample, a red (620nm) colour was obtained. In the case of the red polymorph, the emission peaks of gently and heavily ground samples became 566 and 618nm, respectively. The colour-changes of the orange and red emitters under mild grinding were attributed to crystalline-to-crystalline transformations which alter the molecular configuration. The colour-changes of the yellow and orange emitters under heavy grinding were attributed to transformations from crystalline to amorphous states.

Miscellaneous Mechanochromic Materials

Luminescent compounds based upon a cyanostyrene unit have been developed[287] which exhibited aggregation-induced red, orange, yellow and green emissions. Two of the luminogens exhibited a reversible tricolour-changing mechanochromic luminescence. Two others exhibited a bicolour switching mechanochromism. The bright yellow and orange luminescence colours of two crystalline polymorphs were attributed to their differing molecular packings.

Aminobenzopyranoxanthene exhibits[288,289] switchable near-infrared and blue fluorescences, in which the former is attributed to fluorescence from slip-stacked dimeric structures in crystals, while the latter is attributed to fluorescence from the monomer. The switching between them involves dynamic structural interconversion between the two molecular packing arrangements during grinding and vapour-fuming.

Films composed of ethylcellulose/poly(acrylic acid) exhibit[290] mechanochromic properties over the visible-light range under compression at 130C; some 10C higher than the glass transition temperature. The compressed films largely recover their shape and colour during heat treatment (130C, 10s). The compressive stresses invert the cholesteric helices of the ethylcellulose/poly(acrylic acid) films. It is notable that compressive stresses can alter the handedness of circularly polarized light as well as the wavelength of reflected light.

A mechanochromic colour-change mediated by conformational changes occurring between the folded and twisted conformers of fluorenylidene-acridane compounds with N-alkyl (methyl, ethyl, n-butyl, n-octyl) groups has been observed[291]. The colour, and conformation-change, of a given compound depended upon the nature of the substituent inserted at the nitrogen atom. It also depended upon whether the sample was a solution or a solid, and whether it was crystalline or amorphous. Grinding of N-methyl samples changed the material from crystalline to amorphous, and provoked a conformational change from folded to twisted, plus a mechanochromic colour-change from yellow to dark green (table 11). The colour and conformational changes were reversed by exposure to chloroform.

Iso-indolinone based charge-transfer luminogens which exhibit aggregation-induced emission have been prepared[292] by C-H bond activation. Slight variation of the donor component strongly affected the molecular packing, and thus the stimulus-response behavior. The flexing and twisting of donor components created a loosely bound so-called herring-bone packing which permitted a reversible transformation to occur under multiple stimuli. Modified derivatives of the donor could have a very different packing mode and therefore did not exhibit mechanochromism. It was shown that non-covalent C-

H$\cdots\pi$ and $\pi\cdots\pi$ interactions were essential to imparting mechanochromism under an external force.

By combining a large conjugation core and peripheral phenyl rings, phenyltolyldibenzofulvene was synthesized[293]; a luminogen which exhibits aggregation-induced and crystallization-enhanced emissions. It can form blue (465nm) and blue-green (485nm) emitting crystals, while the amorphous solid emits dark yellow-green (525nm) light. The material's propeller-like conformation leads to a loose packing and easy morphology modification in the solid state. The emission can be reversibly switched between blue, blue-green and yellow-green by mechanical modification of the morphology, or by heating or solvent fuming.

Table 12. Mechanical properties of boron difluoride dibenzoylmethane derivatives

Derivative	E(GPa)	H(MPa)
BF$_2$dbm(Bu)$_2$	0.369	92.45
BF$_2$dbm(OMe)$_2$	10.864	264.93
BF$_2$dbmOMe	8.620	255.79

Crystals of the boron difluoride dibenzoylmethane derivative, BF$_2$dbm(Bu)$_2$ (table 12), can be bent permanently, while the BF$_2$dbm(OMe)$_2$ derivative undergoes inhomogeneous shear deformation. The BF$_2$dbmOMe crystals are brittle. Marked mechanochromic luminescence was observed in the case of BF$_2$dbm(Bu)$_2$, whereas BF$_2$dbm(OMe)$_2$ exhibited only slight mechanochromic luminescence and BF$_2$dbmOMe exhibited no detectable mechanochromic luminescence. It was concluded[294] that mechanochromic luminescence in solid-state organic fluorophores is related to the extent of plasticity.

The material, (E)-5-methoxy-2-(((6-methylbenzo[d]thiazol-2-yl)imino)methyl)phenol[295], exhibits an unique mechanochromic fluorescence in that the effect of mechanical treatment does not appear immediately, but only following fuming.

Vinamidinium with various aryl substitutes, phenyl, p-methylphenyl, p-bromophenyl, p-nitrophenyl and naphthyl, was synthesized[296]. All of the compounds exhibited intense 1π-π*/intramolecular charge-transfer absorption bands between 300 and 450nm. The phenyl and p-bromophenyl derivatives exhibited yellow and orange fluorescences, respectively, and a mechanochromism such that, following grinding, their emissions changed to green–yellow and yellow, respectively. These changes were reversed when the ground powders were fumed, and the reversible mechanochromism was attributed to crystalline-amorphous transformations.

Fluorescent 1,4-diketo-2,5-dibutyl-3,6-bis(4-(carbazol-N-yl)phenyl)pyrrolo[3,4-c]pyrrole was synthesized[297] and exhibited two-photon excitation fluorescence. Both yellow and red crystals, having weak intermolecular interactions, could be ground to give an amorphous orange state and then reverted to yellow by heat-treatment and solvent-fuming. X-ray diffraction and differential scanning calorimetry indicated that transitions between the crystalline and amorphous state explained the mechanochromic behavior.

Red fluorescent materials having the same 1,1,2,2-tetra(thiophen-2-yl)ethene core unit but various branched terminal groups (triphenylamine, N,N-diphenylthiophen-2-amine, N,N-diphenyl-4-vinylaniline) were synthesized[298]. The absorption and emission spectra of the N,N-diphenylthiophen-2-amine and N,N-diphenyl-4-vinylaniline derivatives exhibited red-shifts upon replacing triphenylamine units with N,N-diphenyl-4-vinylaniline and N,N-diphenylthiophen-2-amine terminal groups. This was attributed to increased p-conjugation of the molecule and to more extended skeletons as compared with triphenylamine. The latter exhibited a marked mechanochromic behaviour, with a change from yellow to deep-red in powders, and large (110nm) red-shifts of the fluorescence peaks due to grinding. The N,N-diphenylthiophen-2-amine derivative exhibited weak mechanochromism, with a slight red-shift during grinding. The mechanochromism was attributed to a morphology change from the looser crystalline state, of triphenylamine and N,N-diphenylthiophen-2-amine, to the more compact amorphous state.

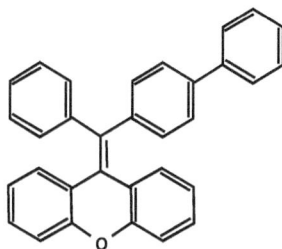

Figure 33. Structure of xanthone-4-benzoylbiphenyl luminogen

A 2,5-dicarbazole-substituted terephthalate derivative consisting of two carboxylic acid groups was synthesized[299] and exhibited a mechanofluorochromic behavior such that, under mechanical treatment, the white solid became yellow and its cyan fluorescence changed to yellow-green. The colour and fluorescence were recoverable by fuming or annealing. When exposed to NH_3 vapor, the pristine and ground solids emitted a blue fluorescence. The fluorescence colours were restored to cyan and yellow-green by fuming with HCOOH vapor.

Donor-acceptor phenothiazine derivatives, functionalized by formyl groups having N-alkyl chains of various lengths, were synthesized[300]. The materials exhibited a reversible mechanochromic behavior and chain-length dependent emissions. Samples with N = 1 exhibited smaller (22nm) fluorescence spectrum shifts under imposed forces. Those having longer alkyl chains exhibited similar mechanochromic behaviors but a greater fluorescence contrast following grinding; apart from that with N = 6. The fluorescence of ground solids with N = 1 or 4 recovered at room temperature. Material with N = 2 required high temperatures for fluorescence recovery.

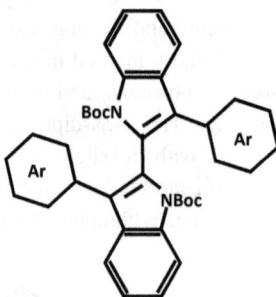

Figure 34. Structures of N,N'-bis-Boc-3,3'-diaryl-2,2'-bi-indole derivatives.
Here, Boc is tert-butoxycarbonyl and Ar can be phenyl, 2-naphthyl or 1-pyrenyl

The luminogen, (E)-3-chloro-4-((4-(diethylamino)-2-hydroxy benzylidene)-amino)benzoic acid is a functionalized Schiff base with a twisted molecular configuration[301], and incorporated electron donor-acceptor pairs which impart aggregation-induced emission and twisted intramolecular charge-transfer. The crystalline material exhibits a strong green (514nm) emission which becomes orange-red (562nm) as a result of grinding. This can be reversed by recrystallization via immersion or fuming in organic solvents; a cycle which can be repeated more than 30 times.

Table 13. Mechanical properties of difluoroboron avobenzone polymorphs

Polymorph	Structure	Orientation	E(GPa)	H(MPa)
Green	orthorhombic	(011)	10.6	275
Green	orthorhombic	(001)	7.4	340
Green	orthorhombic	(T20)	13.2	410
Cyan	monoclinic	(001)	5.6	206

Reacting xanthone with 4-benzoylbiphenyl produces[302] a mechanochromic luminogen which features a large conjugation core plus peripheral phenyl rings (figure 33). It exhibits aggregation-induced and crystallization-enhanced emissions. In particular, the emissions can be switched between deep-blue (432nm), green (492nm) and orange (584nm) by heating, by mechanical treatment and by solvent fuming.

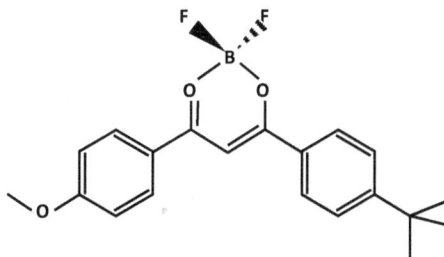

Figure 35. Chemical structure of difluoroboron avobenzone

Fluorenyl-containing tetra-substituted ethylene compounds exhibit[303] a morphology-dependent luminescence, such that the emission colour can be interchanged between green and blue by grinding and solvent fuming, due to the transformation between crystalline and amorphous states. This in turn was suggested to be facilitated by the twisted molecular configurations and loose packing of the crystal.

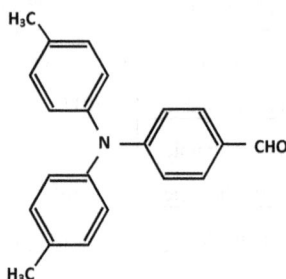

Figure 36. Structure of 4-[bis(4-methylphenyl)amino]benzaldehyde

The emission properties of three N,N'-bis-tert-butoxycarbonyl-3,3'-diaryl-2,2'-bi-indole derivatives (figure 34) as solid-state fluorophores have been studied[304]. In the case of the crystalline 3,3'-dipyrenyl derivative, a partial intramolecular stacking of the dipyrenyl groups was observed and that resulted in a two-step mechanochromic luminescence which went from blue (462nm) to green (516nm) and yellow (535nm).

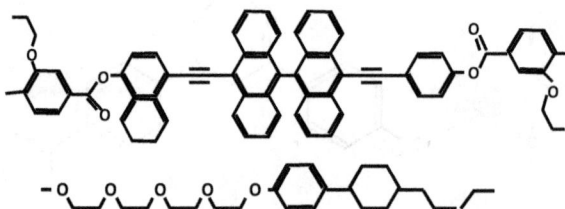

Figure 37. 10,10'-bis(phenylethynyl)-9,9'-bianthryl,
and one of four mesogenic components

Difluoroboron avobenzone, a simple boron complex of a commercial sunscreen, exhibits a morphology-dependent mechanochromic solid-state luminescence[305]. When disturbed to the slightest degree, the emission colour of film samples is appreciably red-shifted under ultra-violet light. The fluorescence of damaged regions recovers slowly at room temperature, but much more rapidly upon heating. A nano-indentation study[306] of the mechanical properties (table 13) of green- and cyan-emitting polymorphs of difluoroboron avobenzone (figure 35) showed that shearing deformation of the softer cyan form provoked a colour-change to yellow, while the harder green form suffered a barely noticeable red-shift. The presence of slip-planes in the cyan form facilitated the formation of recoverable low-energy defects in the structure, thus facilitating reversible mechanochromism.

4-[bis(4-methylphenyl)amino]benzaldehyde (figure 36) exhibited[307] mechanochromic luminescence, in that the fluorescence colour of crystalline samples changed from light-blue to greenish-yellow during grinding. It then returned to the original colour during annealing at room temperature. This behaviour was suggested to be due to the formation of a supercooled liquid state by grinding and recrystallization.

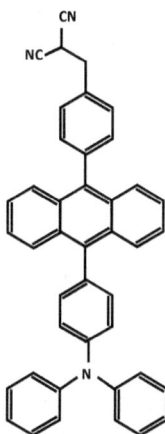

Figure 38. Structure of the triphenylamino-
(2,2-dicyanovinyl)phenyl-anthryl compound

Smectic liquid crystals which comprised an aromatic component, based upon 10,10'-bis(phenylethynyl)-9,9'-bianthryl at the centre, and mesogenic components on each side – linked by oxyethylene spacers - have been investigated[308]. A compound with two mesogenic components on each side (figure 37) formed an ordered smectic phase which exhibited mechanochromic luminescence.

Samples which were formed by annealing emitted a green photoluminescence at room temperature. The photoluminescent colour changed from green to blue-green during mechanical stimulation at room temperature. A change in this colour following mechanical treatment was attributed to a disturbance of the π-π interactions between adjacent molecules. The green photoluminescence was restored by heating, followed by cooling to room temperature. Smaller mesogenic components led to stronger ground-state interactions between adjacent luminescent molecules and to red-shifted emissions. Materials which comprised three mesogenic components on each side exhibited no mechanochromic luminescence. Large mesogenic components disrupted the intermolecular interactions and resulted in no detectable change occurring in the assembly of luminescent components during mechanical treatment.

A mechanochromic luminescent D-π-A compound (figure 38) was constructed[309] by combining a triphenylamino donor with a (2,2-dicyanovinyl)phenyl acceptor; together with an anthryl π-spacer. Monocrystalline samples exhibited a green (532nm) emission but, upon grinding them, there was a sharp colour-change from green to orange-red (615nm). This could be reversed by annealing or solvent-fuming. The emission colour-change during grinding was attributed to partial conversion of the ordered solid-state structure into an amorphous form.

The compound, (E)-6-((4-(diethylamino)-2-hydroxybenzylidene)amino)-2-naphthoic acid, (figure 39) has a twisted configuration which incorporates[310] electron-donor and acceptor pairs. These features impart to it both twisted intramolecular charge transfer and aggregation-induced emission properties. The crystalline powder exhibits a strong green emission, and this changes to orange-red upon grinding. This can be reversed by recrystallization via immersion or fuming in organic solvents; a cycle which can be repeated more than 10 times. In mechanical tests, film samples were subjected to imposed perpendicular pressures by using an indenter with a diameter of 1.5mm. The film started to suffer reduced emissivity, from the force-bearing area, at pressures of up to 67.9MPa.

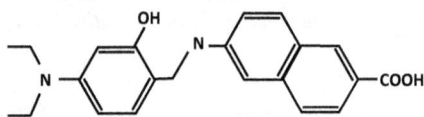

Figure 39. (E)-6-((4-(diethylamino)-2-hydroxybenzylidene)amino)-2-naphthoic acid

Temperature can have a marked effect upon the mechanically-induced behaviour of mechanoresponsive luminescent compounds. The photoluminescence of a cyano-substituted oligo(p-phenylenevinylene) derivative, for example, underwent an appreciable red-shift when it was ground at room temperature[311]. The same treatment, when performed at 100C, led to a hypsochromic shift. The room-temperature behaviour was attributed to a mechanically-induced transition from crystalline to amorphous in which the molecules formed excimers, The hypsochromic shift at high temperatures was attributed to a crystalline-crystalline phase transition.

When 3 series of D-π-A 2,6-dimethyl-4-pyrone-cored derivatives were synthesized[312], spectroscopy showed that all of the compounds exhibited the aggregation-induced emission enhancement effect. It was deduced that the restriction of intramolecular rotation was the key factor in the appearance of aggregation-induced emission enhancement. All of the compounds exhibited a red-shifted mechanochromic behavior (table 14) and the degree of the red-shift depended mainly upon the position of, and length of, the alkoxyl group. Grinding caused a transformation from crystalline to amorphous.

Table 14. Mechanochromic behaviour of 2,6-dimethyl-4-pyrone derivatives

Derivative	Condition	λ(nm)
2,6-bis((E)-2-butoxystyryl)-4H-pyran-4-one	as-received	466
2,6-bis((E)-2-butoxystyryl)-4H-pyran-4-one	pressed	478
2,6-bis((E)-2-butoxystyryl)-4H-pyran-4-one	annealed	464
2,6-bis((E)-2-(octyloxy)styryl)-4H-pyran-4-one	as-received	458
2,6-bis((E)-2-(octyloxy)styryl)-4H-pyran-4-one	pressed	478
2,6-bis((E)-2-(octyloxy)styryl)-4H-pyran-4-one	annealed	458
2,6-bis((E)-2-(dodecyloxy)styryl)-4H-pyran-4-one	as-received	455
2,6-bis((E)-2-(dodecyloxy)styryl)-4H-pyran-4-one	pressed	472
2,6-bis((E)-2-(dodecyloxy)styryl)-4H-pyran-4-one	annealed	460
2,6-bis((E)-2-(hexadecyloxy)styryl)-4H-pyran-4-one	as-received	442
2,6-bis((E)-2-(hexadecyloxy)styryl)-4H-pyran-4-one	pressed	473
2,6-bis((E)-2-(hexadecyloxy)styryl)-4H-pyran-4-one	annealed	465
2,6-bis((E)-4-butoxystyryl)-4H-pyran-4-one	as-received	490
2,6-bis((E)-4-butoxystyryl)-4H-pyran-4-one	pressed	508
2,6-bis((E)-4-butoxystyryl)-4H-pyran-4-one	annealed	493
2,6-bis((E)-4-(octyloxy)styryl)-4H-pyran-4-one	as-received	460

2,6-bis((E)-4-(octyloxy)styryl)-4H-pyran-4-one	pressed	500
2,6-bis((E)-4-(octyloxy)styryl)-4H-pyran-4-one	annealed	481
2,6-bis((E)-2-(dodecyloxy)styryl)-4H-pyran-4-one	as-received	472
2,6-bis((E)-2-(dodecyloxy)styryl)-4H-pyran-4-one	pressed	503
2,6-bis((E)-2-(dodecyloxy)styryl)-4H-pyran-4-one	annealed	480
2,6-bis((E)-2-(naphthalen-1-yl)vinyl)-4H-pyran-4-one	as-received	505
2,6-bis((E)-2-(naphthalen-1-yl)vinyl)-4H-pyran-4-one	pressed	514
2,6-bis((E)-2-(naphthalen-1-yl)vinyl)-4H-pyran-4-one	annealed	504
2,6-bis((E)-2-(2-butoxynaphthalen-1-yl)vinyl)-4H-pyran-4-one	as-received	517
2,6-bis((E)-2-(2-butoxynaphthalen-1-yl)vinyl)-4H-pyran-4-one	pressed	524
2,6-bis((E)-2-(2-butoxynaphthalen-1-yl)vinyl)-4H-pyran-4-one	annealed	513
2,6-bis((E)-2-(2-(octyloxy)naphthalen-1-yl)vinyl)-4Hpyran-4-one	as-received	516
2,6-bis((E)-2-(2-(octyloxy)naphthalen-1-yl)vinyl)-4Hpyran-4-one	pressed	526
2,6-bis((E)-2-(2-(octyloxy)naphthalen-1-yl)vinyl)-4Hpyran-4-one	annealed	512

In very recent work[313], 2 donor-acceptor compounds were synthesized from 2,6-dimethyl-4-pyrone and carbazole compounds. Both products exhibited unique fluorescence properties in the aggregated state. Red-shifts of 102 and 130nm were observed, upon changing the solvent from n-hexane to N,N-dimethylformamide, due to solvatochromism. This phenomenon was attributed to the intramolecular charge transfer effect, as confirmed by density functional theory calculations. Both compounds exhibited an excellent blue-shifted mechanochromic behavior.

Mechanochromic effects, with bathochromic shifts of up to 50nm have been observed[314] following grinding. In two synthesized boroquinols, there was a mechanochromic behavior upon grinding the as-synthesized solid. One compound exhibited a change from 464nm in the as-received state to 514nm in the ground state. This was reversed upon heating at 200C. The colour changed from light-blue for the as-synthesized product to yellow for the ground material. The as-synthesized material possessed a high degree of crystallinity, but this was largely destroyed by grinding. The change in emission wavelength was attributed to the presence of strong intermolecular interactions, such as π-π, in the ordered phase. Shear forces were expected to change the intermolecular interactions at the molecular level and therefore the degree of crystallinity. The resultant metastable state and its accompanying intermolecular interactions then permitted various transitions to occur under excitation, and cause the observed changes in the emission wavelength. A similar mechanochromic effect was observed in another boroquinol, which exhibited a bathochromic shift of +28nm during grinding, but reversibility was not possible here because of decomposition at only 60C.

The oxidation of dicyanomethyl-substituted hexa-peri-hexabenzocoronene yields[315] a tetracyanoethylene-bridged hexa-peri-hexabenzocoronene dimer and trimer. In solution, these hexa-peri-hexabenzocoronene oligomers exhibited a conformational isomerism which depended upon the solvent and temperature. Solid samples of the hexa-peri-hexabenzocoronene dimer and trimer exhibited mechanochromism, with a near-infrared absorption appearing upon grinding, due to the generation of radical species. Such radicals were stable for up to 4 months. In particular, the colour of the dimer changed from yellow to brown during grinding. A drop of dichloromethane, deposited onto the ground powder, restored the original yellow colour and the near-infrared absorption bands disappeared. The mechanochromic behaviour of these hexa-peri-hexabenzocoronene oligomers originated from reversible C-C bond dissociation and from the formation of a tetracyanoethylene linkage between the hexa-peri-hexabenzocoronene units. It was noted that compounds which contain long carbon–carbon single bonds often undergo C–C bond dissociation during heating, pressing or grinding, and the C–C bonds between dicyanomethyl units have a tendency to break and release dicyanomethyl radicals. The trimer exhibited a similar mechanochromic behaviour: the as-received material was yellow but turned brown during grinding. In the solid state, the ground sample exhibited a near-infrared absorption band which disappeared following the addition of a drop of dichloromethane. There was again the formation of a radical species during grinding.

Study of a π-conjugated bola-amphiphilic chromophore: oligo-(p-phenylene ethynylene) dicarboxylic acid with dialkoxyoctadecyl side-chains revealed[316] that polymorphs at 123 and 373K also exhibited thermotropic liquid crystallinity and a columnar phase. The inherent π-conjugation of one of the materials imparted luminescence to the system, and photoluminescence measurements of the mesophase revealed a similar luminescence to that of the crystalline state. A mechano-hypsochromic luminescence behavior was also detected. Nano-indentation testing of single crystals revealed a Young's modulus of 5.63GPa and a hardness of 400MPa for the (001) surface and corresponding values of 4.74GPa and 171MPa for the (043) face; thus indicating that the mechanical response, when measured on (001), was much stiffer and harder. The figures for (043) indicated the presence of favorably-oriented slip-planes, allowing relatively easy shear sliding of the molecular planes. Such nano-indentation data helped to explain the mechanoluminescence behaviour. It was hypothesized that the movement of slip planes containing the weaker C−H···π, C−H···O and van der Waals interactions were responsible for the phenomenon. Upon applying pressure, the energy was sufficient to break the weak interactions and a plane could then glide, giving rise to mechanoluminescence. It was possible that the hypsochromic shift was due to monomeric and oligomeric contributions. This created a new phase having a high defect-density. The elastic energy at this point was however insufficient to restore the crystal to its original state, causing an irreversible change in the luminescence. This non-reversibility was further exacerbated by the large residual penetration depth (250nm) after unloading so that, upon removing the force, the planes were slow to resume their original position. It was concluded that nano-indentation data offered a quantitative explanation for liquid crystallinity and for irreversible mechanoluminescence. Both ultra-violet and photoluminescence profiles had confirmed the occurrence of mechanochromic luminescence. The ultra-violet profiles of ground crystals were more broadened. The absorption maximum also appeared to be blue-shifted by 7nm; indicating a change in the absorption behavior during grinding. Photoluminescence profiles also confirmed the mechanoluminescence change. The emission profile of ground film samples had a broadened profile which included the blue and cyan portions of spectrum, with a hypsochromic shift of 5nm. It was proposed that, during grinding, the supramolecular organization of molecular fragments gave rise to monomeric and oligomeric states. Their contributions then gave rise to shifts and to broadening of the spectrum. The behavior was checked for reversibility by annealing for one hour at room temperature or at 60C. In neither case was a reversion to the original emission colour observed.

Blends of polyamide containing small (0.15 to 1wt%) amounts of excimer-forming fluorescent dye, 1,4-bis(α-cyano-4-octadecyloxystyryl)-2,5-dimethoxybenzene, have been produced by melt-processing[317]. A green monomer fluorescence arising from individual chromophores was observed at low concentrations (0.15wt%), but higher concentrations led to aggregation of the dye and the emission was then dominated by a red excimer fluorescence. Deformation of samples with a dye content of 0.25wt% produced a marked mechanochromic effect: a transformation from excimer-dominated to monomer-rich emission. The monomer/excimer emission-ratio could be doubled, in a step-wise manner, when samples were uniaxially deformed to above the yield point (figure 40).

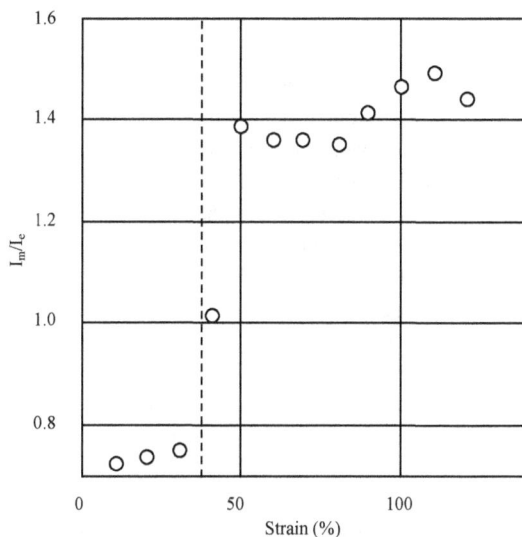

*Figure 40. Ratio of monomer to excimer emission intensity as a function
of uniaxial strain for a of polyamide blend film. Dashed line indicates the yield point*

It has been noted[318] that (E)-4-(((2-hydroxynaphthalen-1-yl)methylene)amino)benzoic acid exhibits aggregation-induced emission and intramolecular charge transfer. Crystalline powder samples emit an intense green fluorescence under ultra-violet illumination but, following grinding, the emission sharply decreases; with a red-shift of the emission wavelength. The changes in fluorescence were attributed to a crystalline-amorphous transition. The emission of the amorphous phase could be returned to its original state by recrystallization via immersion or fuming in an organic solvent. This cycle could be repeated 25 times without fatigue.

A luminescent dye, 2-(m-trifluoromethylanilino)-N-(m-trifluoromethylphenyl)-maleimide, based upon a simple aminomaleimide skeleton was synthesized[319] which exhibited an on/off mechanochromic luminescence. A green emission was produced by grinding, and was turned off by heating or by fuming with dichloromethane. In the crystals, two molecules were stacked by co-facial π-π interactions which provoked concentration self-quenching. The crystalline-amorphous transition which was caused by grinding acted so as to remove co-facial π-π stacking, leading to intense emissions. Crystallization recovered the co-facial π-π stacking and resulted in removal of the emission. Both theory and X-ray diffraction indicated that the dye molecule was distorted in crystals.

The so-called fluorescent-phosphorescent dual-emission material, 4-iodo-4'-(N-carbazolyl) diphenylsulfone, was synthesized[320] in an attempt to enhance the phosphorescence of 4-(N-carbazolyl)diphenylsulfone by exploiting the internal heavy atom effect. It had been found that substituting a single iodine atom at the para position of the free phenyl group in 4-(N-carbazolyl)diphenylsulfone had only a small effect upon the molecular energy levels. Crystals of 4-(N-carbazolyl)diphenylsulfone possessed excellent fluorescent-phosphorescent dual-emission characteristics, with large separate dual emissions. The fluorescent and phosphorescent peaks were located at 383 and 559nm. The phosphorescent peak was enhanced, even to above the fluorescent peak, due to the heavy-atom effect of iodine. The crystalline 4-iodo-4'-(N-carbazolyl) diphenylsulfone, with very separate dual emissions, reacted sensitively to grinding. During grinding, the fluorescent peak at 383nm gradually decreased and a new peak appeared at 412nm. The phosphorescent peak at 559nm steadily reduced but was located at the same position. The photoluminescence quantum-yield also decreased continuously as the phosphorescent part rapidly decreased. Following grinding for 120s, the peaks at 383 and 559nm decreased and the new peak at 412nm became the most intense one. The result of grinding was therefore analogous to the mixing of two colours. Two very different colours were observed: orange from the original peaks, and purple from the new peak. Thus a wide range of emissions, from orange to purple, could occur during grinding; unlike the overall red-shift which was usually observed among other mechanochromic materials. The emission color of ground samples arose from the colour-mixing of orange and purple in various ratios. It was possible to adjust the emission color to white by suitably grinding crystalline samples according to the principles of color-mixing. The amorphous state could be completely restored to crystalline by exposure to dichloromethane vapor.

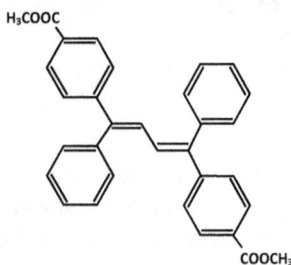

*Figure 41. Structure of dimethyl 4,4'-((1Z,3Z)-1,4-diphenylbuta
-1,3-diene-1,4-diyl)dibenzoate*

Three dyes which were based upon N-phenylcarbazol-substituted tetra-arylethene, were synthesized[321] in order to gauge the effect of substituent groups, in the tetra-arylethene, upon mechanochromism. All of the luminogens exhibited typical aggregation-induced emission characteristics. Only one of them exhibited notable mechanochromism, and grinding–fuming cycles could be repeated many times without fatigue. This indicated that the introduction of hydrophilic groups into hydrophobic tetra-arylethene could change the molecule-packing in the solid state. The emission behaviour was examined by grinding. As-received powder of the above material emitted sky-blue (441nm) light. Upon grinding, the colour changed to cyan, with a maximum emission at 505nm. Luminogens which exhibit mechanochromism can normally be restored by annealing or fuming. However, even after heating at 60, 80, 100 or 120C for more than five hours, the cyan colour could not be changed. Upon fuming with dichloromethane or ethyl acetate vapour for 180s, the original blue colour was recovered. The cycling between blue and green emissions could be repeated without fatigue, due to the non-destructive nature of the mechanical treatment.

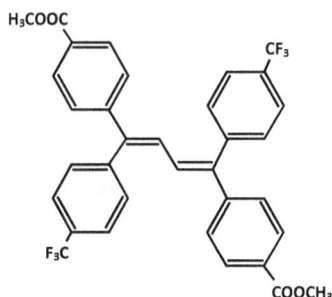

Figure 42. Structure of dimethyl 4,4'-((1Z,3Z)-1,4-bis(4-(trifluoromethyl)phenyl) buta-1,3-diene-1,4-diyl)dibenzoate

Three tetra-aryl substituted 1,3-butadiene derivatives which exhibited aggregation-enhanced emission and mechanochromic fluorescence were synthesized[322] which involved D-π-A structures having large dipole moments. The derivatives contained differing electronic push and/or pull substituents. The fluorescence of as-received dimethyl 4,4'-((1Z,3Z)-1,4-bis(4-(diphenylamino) phenyl)buta-1,3-diene-1,4-diyl)dibenzoate powder changed from green to yellow during irradiation with ultra-violet light or during grinding.

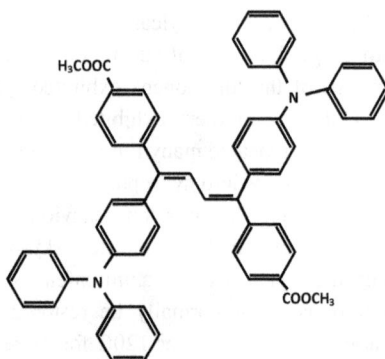

*Figure 43. Structure of dimethyl 4,4'-((1Z,3Z)-1,4-bis(4-(diphenylamino)
phenyl)buta-1,3-diene-1,4-diyl)dibenzoate*

The other materials also exhibited a pressure-induced luminescence change, but the variation of the emission change of dimethyl 4,4'-((1Z,3Z)-1,4-diphenylbuta-1,3-diene-1,4-diyl)dibenzoate (figure 41) and dimethyl 4,4'-((1Z,3Z)-1,4-bis(4-(trifluoromethyl)phenyl)buta-1,3-diene-1,4-diyl)dibenzoate was not as significant. During grinding, the fluorescence colours gradually red-shifted from sky-blue to cyan in the former material and from dark-blue to cyan in the latter material. The maximum emission wavelengths for as-received and ground samples of dimethyl 4,4'-((1Z,3Z)-1,4-bis(4-(trifluoromethyl)phenyl)buta-1,3-diene-1,4-diyl)dibenzoate (figure 42), dimethyl 4,4'-((1Z,3Z)-1,4-diphenylbuta-1,3-diene-1,4-diyl)dibenzoate and dimethyl 4,4'-((1Z,3Z)-1,4-bis(4-(diphenylamino) phenyl)buta-1,3-diene-1,4-diyl)dibenzoate (figure 43) shifted from 459, 455 and 520nm to 465, 465 and 540nm, respectively. The emission colours of the ground samples could be restored almost to their original colours by adding a drop of polar solvent or by heating (180C, 2h). The cycles could be repeated many times, almost without deterioration. The mechanochromism of these materials was attributed to disruption of the molecular packing. It was concluded that the electronic structures of the materials exerted a large effect upon their mechanochromism. Overall, it was proposed that the differences in the mechanochromic behaviors of the three compounds was due to the varied extents of push–pull interaction involved. The substitution of electron-donating and/or electron-accepting groups increased the molecular polarity and led to the marked influence of intermolecular interactions in the crystalline and less-ordered states of the materials. The existence of a large dipole moment is favorable to the appearance of

Materials Research Forum LLC

https://doi.org/10.21741/9781644900277

appreciable mechanochromism. The basic material, shorn of any substituents was synthesized, showing that it was also an aggregation-enhanced emission luminogen. Its maximum emission wavelength for as-received and ground samples of high symmetry was shifted only from 427 to 428nm. The fluorescence of the as-received powder remained dark-blue under ultra-violet light and during grinding. As expected therefore, the basic material exhibited negligible mechanochromism.

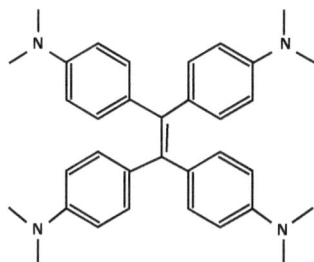

Figure 44. Structure of tetrakis(4-(dimethylamino)phenyl)ethylene

The material, tetrakis(4-(dimethylamino)phenyl)ethylene, with its natural propeller shape and almost centrosymmetric structure (figure 44) has been investigated[323]. A considerable change in the visible colour and UV/vis absorption band was observed before and after grinding. The maximum absorption band of ground samples exhibited a clear red-shift, while the fluorescence of as-received samples could also change from blue (470nm) to yellow-green (528nm) during grinding. The red-shifts of the absorption and fluorescence spectra due to grinding were suggested to arise from increased conjugation of molecules in the ground sample. The maximum absorption band and fluorescence peak were almost restored by fuming with ethyl acetate. Following the annealing (150C, 60s) of ground samples, the emission changed from yellow-green (528nm) to green (492nm); differing from the change caused by fuming. It appeared that the sample which formed during grinding did not return to the as-received blue emission form upon heating, but transformed to a green-emitting sample. The molecules in blue crystals had an asymmetrical configuration while those in green crystals had a symmetrical one. Due to the twisted structure of tetraphenylethylene, the molecules of both crystals were separate from each other and the crystals were thus arranged mainly by the intermolecular interactions between the peripheral units. The stacking prevented molecules from coming close and forming intermolecular π−π stackings, thus having a marked effect upon the emissions because excimers formed due to π−π stacking usually have a red-shifted emission, as compared to that of a monomer. At the same time, the interactions between the transition dipoles of adjacent molecules were weak due to the large intermolecular distances. It was concluded that the emission red-shift following grinding depended mainly upon the intramolecular conformation.

An attempt[324] was made to elucidate the mechanism of difluoroboron dibenzoylmethane mechanochromism by investigating excited-state interactions among the molecules, and how those interactions change during mechanical solicitation. Upon correlating solution and solid-state data, it was deduced that the coupled processes of force-induced emissive H-aggregate formation, and energy-transfer to the emissive H-aggregates, were responsible for the observed mechanoluminescence.

Just recently[325], a novel means has been found for sensitizing the centrosymmetrically-packed organic luminogens which tend not to react to external influences due to their zero overall dipole moment and degenerate solid-state electronic energy states. Fused bi-heterocyclic luminogens with centrosymmetric packing and organic luminogens have been prepared by using a metal-catalyzed C-H bond-activation method. The as-prepared luminogens do not respond to external mechanical stresses, but heating provokes a so-called centrosymmetric to non-centrosymmetric phase transition at the crystallization temperature. Following this phase transition, the luminogens then react to external mechanical influences. It is presumed that the higher dipole moments of luminogens in non-centrosymmetric crystals are responsible for these surprising mechanochromic properties.

References

[1]   Longchambon, H. Bulletin de la Société Française de Minéralogie, 1925, 48, 130-147.

[2]   Sweeting, L.M., Chemistry of Materials, 13[3] 2001, 854-870.

[3]   Calvino, C., Sagara, Y., Buclin, V., Haehnel, A.P., del Prado, A., Aeby, C., Simon, Y.C., Schrettl, S., Weder, C., Macromolecular Rapid Communications, 40[1] 2019, 1800705. https://doi.org/10.1002/marc.201800705

[4]   Jiang, Y., Zeng, S., Yao, Y., Xu, S., Dong, Q., Chen, P., Wang, Z., Zhang, M., Zhu, M., Xu, G., Zeng, H., Sun, L., Polymers, 11[1] 2019, 103.

[5]   Pisarenko, L.M., Gagarina, A.B., Roginskii, V.A., Bulletin of the Academy of Sciences of the USSR Division of Chemical Science, 36[12] 1987, 2658-2660. https://doi.org/10.1007/bf00957263

[6]   Pisarenko, L.M., Nikulin, V.I., Blagorazumov, M.P., Neiland, O.Y., Paulinsh, L.L., Bulletin of the Academy of Sciences of the USSR Division of Chemical Science, 39[7] 1990, 1379-1385.

[7]   Hou, Y., Du, J., Hou, J., Shi, P., Wang, K., Zhang, S., Han, T., Li, Z., Dyes and Pigments, 160, 2019, 830-838.

[8]   Ishizuki, K., Aoki, D., Goseki, R., Otsuka, H., ACS Macro Letters, 7[5] 2018, 556-560. https://doi.org/10.1021/acsmacrolett.8b00224

[9]   Ito, S., Katada, G., Taguchi, T., Kawamura, I., Ubukata, T., Asami, M., CrystEngComm, 21[1] 2019, 53-59. https://doi.org/10.1039/c8ce01698d

[10] Metzler, L., Reichenbach, T., Brügner, O., Komber, H., Lombeck, F., Müllers, S., Hanselmann, R., Hillebrecht, H., Walter, M., Sommer, M., Polymer Chemistry, 6[19] 2015, 3694-3707. https://doi.org/10.1039/c5py00141b

[11] Wong, B.M., Ye, S.H., O'Bryan, G., Nanoscale, 4[4] 2012, 1321-1327.

[12] O'Bryan, G., Wong, B.M., McElhanon, J.R., ACS Applied Materials and Interfaces, 2[6] 2010, 1594-1600.

[13] Peterson, G.I., Larsen, M.B., Ganter, M.A., Storti, D.W., Boydston, A.J., ACS Applied Materials and Interfaces, 7[1] 2015, 577-583. https://doi.org/10.1021/am506745m

[14] Ma, Z., Meng, X., Ji, Y., Li, A., Qi, G., Xu, W., Zou, B., Ma, Y., Kuang, G.C., Jia, X., Dyes and Pigments, 162, 2019, 136-144. https://doi.org/10.1016/j.dyepig.2017.11.036

[15] Mo, S., Tan, L., Fang, B., Wu, Z., Su, Z., Zhang, Y., Yin, M., Science China Chemistry, 61[12] 2018, 1587-1593. https://doi.org/10.1007/s11426-018-9303-9

[16] Lin, Y., Barbee, M.H., Chang, C.C., Craig, S.L., Journal of the American Chemical Society, 140[46] 2018, 15969-15975. https://doi.org/10.1021/jacs.8b10376

[17] Huang, L., Wu, C., Zhang, L., Ma, Z., Jia, X., ACS Applied Materials and Interfaces, 10[40] 2018, 34475-34484.

[18] Raisch, M., Genovese, D., Zaccheroni, N., Schmidt, S.B., Focarete, M.L., Sommer, M., Gualandi, C., Advanced Materials, 30[39] 2018, 1802813. https://doi.org/10.1002/adma.201802813

[19] Barbee, M.H., Mondal, K., Deng, J.Z., Bharambe, V., Neumann, T.V., Adams, J.J., Boechler, N., Dickey, M.D., Craig, S.L., ACS Applied Materials and

Interfaces, 10[35] 2018, 29918-29924. https://doi.org/10.1021/acsami.8b09130

[20]   Kempe, F., Brügner, O., Buchheit, H., Momm, S.N., Riehle, F., Hameury, S., Walter, M., Sommer, M., Angewandte Chemie, 57[4] 2018, 997-1000. https://doi.org/10.1002/ange.201709142

[21]   Shree, S., Schulz-Senft, M., Alsleben, N.H., Mishra, Y.K., Staubitz, A., Adelung, R., ACS Applied Materials and Interfaces, 9[43] 2017, 38000-38007. https://doi.org/10.1021/acsami.7b09598

[22]   Meng, X., Chen, C., Qi, G., Li, X., Wang, K., Zou, B., Ma, Y., ChemNanoMat, 3[8] 2017, 569-574.

[23]   Li, M., Liu, W., Zhang, Q., Zhu, S., ACS Applied Materials and Interfaces, 9[17] 2017, 15156-15163.

[24]   Su, X., Wang, Y., Fang, X., Zhang, Y.M., Zhang, T., Li, M., Liu, Y., Lin, T., Zhang, S.X.A., Chemistry - an Asian Journal, 11[22] 2016, 3205-3212.

[25]   Wan, S., Ma, Z., Chen, C., Li, F., Wang, F., Jia, X., Yang, W., Yin, M., Advanced Functional Materials, 26[3] 2016, 353-364.

[26]   Meng, X., Qi, G., Li, X., Wang, Z., Wang, K., Zou, B., Ma, Y., Journal of Materials Chemistry C, 4[32] 2016, 7584-7588.

[27]   Meng, X., Qi, G., Zhang, C., Wang, K., Zou, B., Ma, Y., Chemical Communications, 51[45] 2015, 9320-9323.

[28]   Hemmer, J.R., Smith, P.D., Van Horn, M., Alnemrat, S., Mason, B.P., De Alaniz, J.R., Osswald, S., Hooper, J.P., Journal of Polymer Science B, 52[20] 2014, 1347-1356. https://doi.org/10.1002/polb.23569

[29]   Van Horn, M., Smith, P., Mason, B.P., Hemmer, J.R., Read De Alaniz, J., Hooper, J.P., Osswald, S., Journal of Applied Physics, 117[4] 2015, 043103. https://doi.org/10.1063/1.4906326

[30]   Fang, X., Zhang, H., Chen, Y., Lin, Y., Xu, Y., Weng, W., Macromolecules, 46[16] 2013, 6566-6574.

[31]   Chen, Y., Zhang, H., Fang, X., Lin, Y., Xu, Y., Weng, W., ACS Macro Letters, 3[2] 2014, 141-145.

[32]   Zhang, H., Chen, Y., Lin, Y., Fang, X., Xu, Y., Ruan, Y., Weng, W., Macromolecules, 47[19] 2014, 6783-6790. https://doi.org/10.1021/ma500760p

[33]   Qiu, W., Gurr, P.A., Da Silva, G., Qiao, G.G., Polymer Chemistry, 10[13] 2019,

1650-1659.

[34] Osawa, M., Kawata, I., Igawa, S., Hoshino, M., Fukunaga, T., Hashizume, D., Chemistry - a European Journal, 16[40] 2010, 12114-12126. https://doi.org/10.1002/chem.201001908

[35] Jobbágy, C., Deák, A., European Journal of Inorganic Chemistry, 27, 2014, 4434-4449.

[36] Seki, T., Kashiyama, K., Yagai, S., Ito, H., Chemistry Letters, 46[9] 2017, 1415-1418.

[37] Wong, B.Y.W., Wong, H.L., Wong, Y.C., Au, V.K.M., Chan, M.Y., Yam, V.W.W., Chemical Science, 8[10] 2017, 6936-6946.

[38] Wang, J., He, Q., Wang, C., Bu, W., Soft Matter, 14[1] 2017, 31-34.

[39] Seki, T., Takamatsu, Y., Ito, H., Journal of the American Chemical Society, 138[19] 2016, 6252-6260.

[40] Seki, T., Ozaki, T., Okura, T., Asakura, K., Sakon, A., Uekusa, H., Ito, H., Chemical Science, 6[4] 2015, 2187-2195.

[41] Liu, Q., Xie, M., Chang, X., Gao, Q., Chen, Y., Lu, W., Chemical Communications, 54[91] 2018, 12844-12847.

[42] Wang, X.Y., Zhang, J., Dong, Y.B., Zhang, Y., Yin, J., Liu, S.H., Dyes and Pigments, 156, 2018, 74-81.

[43] Seki, T., Sakurada, K., Ito, H., Chemical Communications, 51[73] 2015, 13933-13936.

[44] Walters, D.T., Aghakhanpour, R.B., Powers, X.B., Ghiassi, K.B., Olmstead, M.M., Balch, A.L., Journal of the American Chemical Society, 140[24] 2018, 7533-7542. https://doi.org/10.1021/jacs.8b01666

[45] Li, W.B., Luo, W.J., Li, K.X., Yuan, W.Z., Zhang, Y.M., Chinese Chemical Letters, 28[6] 2017, 1300-1305.

[46] Seki, T., Tokodai, N., Omagari, S., Nakanishi, T., Hasegawa, Y., Iwasa, T., Taketsugu, T., Ito, H., Journal of the American Chemical Society, 139[19] 2017, 6514-6517. https://doi.org/10.1021/jacs.7b00587

[47] Yagai, S., Seki, T., Aonuma, H., Kawaguchi, K., Karatsu, T., Okura, T., Sakon, A., Uekusa, H., Ito, H., Chemistry of Materials, 28[1] 2016, 234-241. https://doi.org/10.1021/acs.chemmater.5b03932

[48]   Qian, Z., Deng, W., Zhang, X., Miao, H., Zhang, G., RSC Advances, 7[74] 2017, 46721-46725. https://doi.org/10.1039/c7ra09453a

[49]   Chen, K., Nenzel, M.M., Brown, T.M., Catalano, V.J., Inorganic Chemistry, 54[14] 2015, 6900-6909.

[50]   Liang, J., Hu, F., Lv, X., Chen, Z., Chen, Z., Yin, J., Yu, G.A., Liu, S.H., Dyes and Pigments, 95[3] 2012, 485-490.

[51]   Liang, J., Chen, Z., Xu, L., Wang, J., Yin, J., Yu, G.A., Chen, Z.N., Liu, S.H., Journal of Materials Chemistry C, 2[12] 2014, 2243-2250.

[52]   Chen, Z., Huang, P.S., Li, Z., Yin, J., Yu, G.A., Liu, S.H., Inorganica Chimica Acta, 432, 2015, 192-197.

[53]   Chen, Z., Yang, L., Hu, Y., Wu, D., Yin, J., Yu, G.A., Liu, S.H., RSC Advances, 5[114] 2015, 93757-93764.

[54]   Xue, S., Qiu, X., Sun, Q., Yang, W., Journal of Materials Chemistry C, 4[8] 2016, 1568-1578.

[55]   Chen, Z., Li, Z., Hu, F., Yu, G.A., Yin, J., Liu, S.H., Dyes and Pigments, 125, 2016, 169-178.

[56]   Chen, Z., Liu, G., Pu, S., Liu, S.H., Dyes and Pigments, 143, 2017, 409-415.

[57]   Chen, Z., Zhang, J., Song, M., Yin, J., Yu, G.A., Liu, S.H., Chemical Communications, 51[2] 2015, 326-329.

[58]   Chen, Z., Nie, Y., Liu, S.H., RSC Advances, 6[77] 2016, 73933-73938.

[59]   Chen, Z., Li, Z., Yang, L., Liang, J., Yin, J., Yu, G.A., Liu, S.H., Dyes and Pigments, 121, 2015, 170-177.

[60]   Dong, Y.B., Chen, Z., Yang, L., Hu, Y.X., Wang, X.Y., Yin, J., Liu, S.H., Dyes and Pigments, 150, 2018, 315-322.

[61]   Chen, Z., Liu, G., Pu, S., Liu, S.H., Dyes and Pigments, 152, 2018, 54-59.

[62]   Chen, Z., Liu, G., Pu, S., Liu, S.H., Dyes and Pigments, 159, 2018, 499-505.

[63]   Chen, Z., Liu, G., Wang, R., Pu, S., RSC Advances, 7[25] 2017, 15112-15115.

[64]   Kawaguchi, K., Seki, T., Karatsu, T., Kitamura, A., Ito, H., Yagai, S., Chemical Communications, 49[97] 2013, 11391-11393.

[65]   Baranyai, P., Marsi, G., Jobbágy, C., Domján, A., Oláh, L., Deák, A., Dalton Transactions, 44[30] 2015, 13455-13459. https://doi.org/10.1039/c5dt01795e

[66] Deák, A., Jobbágy, C., Marsi, G., Molnár, M., Szakács, Z., Baranyai, P., Chemistry - a European Journal, 21[32] 2015, 11495-11508. https://doi.org/10.1002/chem.201501066

[67] Ito, H., Saito, T., Oshima, N., Kitamura, N., Ishizaka, S., Hinatsu, Y., Wakeshima, M., Kato, M., Tsuge, K., Sawamura, M., Journal of the American Chemical Society, 130[31] 2008, 10044-10045. https://doi.org/10.1021/ja8019356

[68] Chen, Z., Liu, G., Wang, R., Pu, S., RSC Advances, 7, 2017, 15112-15115. Photographs abstracted from this reference (DOI: 10.1039/c7ra00913e) under Creative Commons Licence 3.0.

[69] Wu, N.M.W., Ng, M., Yam, V.W.W., Angewandte Chemie, 58[10] 2019, 3027-3031.

[70] Benito, Q., Maurin, I., Cheisson, T., Nocton, G., Fargues, A., Garcia, A., Martineau, C., Gacoin, T., Boilot, J.P., Perruchas, S., Chemistry - a European Journal, 21[15] 2015, 5892-5897. https://doi.org/10.1002/chem.201500251

[71] Benito, Q., Maurin, I., Poggi, M., Martineau-Corcos, C., Gacoin, T., Boilot, J.P., Perruchas, S., Journal of Materials Chemistry C, 4[47] 2016, 11231-11237. https://doi.org/10.1039/c6tc04262g

[72] Lu, T., Wang, J.Y., Tu, D., Chen, Z.N., Chen, X.T., Xue, Z.L., Inorganic Chemistry, 57[21] 2018, 13618-13630.

[73] Kobayashi, A., Yoshida, Y., Yoshida, M., Kato, M., Chemistry - a European Journal, 24[55] 2018, 14750-14759.

[74] Kathewad, N., Pal, S., Kumawat, R.L., Ehesan Ali, M., Khan, S., European Journal of Inorganic Chemistry, 22, 2018, 2518-2523. https://doi.org/10.1002/ejic.201800096

[75] Feng, N., Gao, C., Guo, C.Y., Chen, G., ACS Applied Materials and Interfaces, 10[6] 2018, 5603-5608.

[76] Hu, L.X., Gao, M., Wen, T., Kang, Y., Chen, S., Inorganic Chemistry, 56[11] 2017, 6507-6511.

[77] Kwon, E., Kim, J., Lee, K.Y., Kim, T.H., Inorganic Chemistry, 56[2] 2017, 943-949.

[78] Aguirrechu-Comerón, A., Hernández-Molina, R., Rodríguez-Hernández, P., Muñoz, A., Rodríguez-Mendoza, U.R., Lavín, V., Angel, R.J., Gonzalez-Platas, J., Inorganic Chemistry, 55[15] 2016, 7476-7484.

https://doi.org/10.1021/acs.inorgchem.6b00796

[79] Wen, T., Zhang, D.X., Liu, J., Zhang, H.X., Zhang, J., Chemical Communications, 51[7] 2015, 1353-1355.

[80] Wen, T., Zhang, D.X., Zhang, H.X., Zhang, H.B., Zhang, J., Li, D.S., Chemical Communications, 50[63] 2014, 8754-8756.

[81] Zhang, D.X., Zhang, H.X., Wen, T., Li, D.S., Zhang, J., Inorganic Chemistry Frontiers, 3[2] 2016, 263-267.

[82] Benito, Q., Baptiste, B., Polian, A., Delbes, L., Martinelli, L., Gacoin, T., Boilot, J.P., Perruchas, S., Inorganic Chemistry, 54[20] 2015, 9821-9825. https://doi.org/10.1021/acs.inorgchem.5b01546

[83] Hupp, B., Nitsch, J., Schmitt, T., Bertermann, R., Edkins, K., Hirsch, F., Fischer, I., Auth, M., Sperlich, A., Steffen, A., Angewandte Chemie, 57[41] 2018, 13671-13675. https://doi.org/10.1002/anie.201807768

[84] Wu, Z., Liu, J., Gao, Y., Liu, H., Li, T., Zou, H., Wang, Z., Zhang, K., Wang, Y., Zhang, H., Yang, B., Journal of the American Chemical Society, 137[40] 2015, 12906-12913. https://doi.org/10.1021/jacs.5b06550

[85] Conesa-Egea, J., Nogal, N., Martínez, J.I., Fernández-Moreira, V., Rodríguez-Mendoza, U.R., González-Platas, J., Gómez-García, C.J., Delgado, S., Zamora, F., Amo-Ochoa, P., Chemical Science, 9[41] 2018, 8000-8010. https://doi.org/10.1039/c8sc03085e

[86] Kobayashi, A., Kato, M., Chemistry Letters, 46[2] 2017, 154-162.

[87] Perruchas, S., Goff, X.F.L., Maron, S., Maurin, I., Guillen, F., Garcia, A., Gacoin, T., Boilot, J.P., Journal of the American Chemical Society, 132[32] 2010, 10967-10969. https://doi.org/10.1021/ja103431d

[88] Huitorel, B., El Moll, H., Cordier, M., Fargues, A., Garcia, A., Massuyeau, F., Martineau-Corcos, C., Gacoin, T., Perruchas, S., Inorganic Chemistry, 56[20] 2017, 12379-12388. https://doi.org/10.1021/acs.inorgchem.7b01870

[89] Huitorel, B., Benito, Q., Fargues, A., Garcia, A., Gacoin, T., Boilot, J.P., Perruchas, S., Camerel, F., Chemistry of Materials, 28[22] 2016, 8190-8200. https://doi.org/10.1021/acs.chemmater.6b03002

[90] Benito, Q., Le Goff, X.F., Maron, S., Fargues, A., Garcia, A., Martineau, C., Taulelle, F., Kahlal, S., Gacoin, T., Boilot, J.P., Perruchas, S., Journal of the American Chemical Society, 136[32] 2014, 11311-11320.

https://doi.org/10.1021/ja500247b

[91]   Deshmukh, M.S., Yadav, A., Pant, R., Boomishankar, R., Inorganic Chemistry,
       54[4] 2015, 1337-1345.

[92]   Wen, T., Zhou, X.P., Zhang, D.X., Li, D., Chemistry - a European Journal, 20[3]
       2014, 644-648.

[93]   Lin, C.J., Liu, Y.H., Peng, S.M., Shinmyozu, T., Yang, J.S., Inorganic Chemistry,
       56[9] 2017, 4978-4989.

[94]   Naziruddin, A.R., Lee, C.S., Lin, W.J., Sun, B.J., Chao, K.H., Chang, A.H.H.,
       Hwang, W.S., Dalton Transactions, 45[13] 2016, 5848-5859.

[95]   Henwood, A.F., Webster, J., Cordes, D., Slawin, A.M.Z., Jacquemin, D., Zysman-
       Colman, E., RSC Advances, 7[41] 2017, 25566-25574.
       https://doi.org/10.1039/c7ra03190d

[96]   Choi, S.J., Kuwabara, J., Nishimura, Y., Arai, T., Kanbara, T., Chemistry Letters,
       41[1] 2012, 65-67.

[97]   Zhang, X.P., Zhang, D.S., Qi, X.W., Zhu, L.H., Wang, X.H., Sun, W., Shi, Z.F.,
       Lin, Q., Inorganica Chimica Acta, 467, 2017, 99-105.

[98]   Zhang, X., Zhu, L., Wang, X., Shi, Z., Lin, Q., Inorganica Chimica Acta, 442,
       2016, 56-63.

[99]   Zhang, X.P., Mei, J.F., Lai, J.C., Li, C.H., You, X.Z., Journal of Materials
       Chemistry C, 3[10] 2015, 2350-2357.

[100] Genovese, D., Aliprandi, A., Prasetyanto, E.A., Mauro, M., Hirtz, M., Fuchs, H.,
       Fujita, Y., Uji-I, H., Lebedkin, S., Kappes, M., De Cola, L., Advanced Functional
       Materials, 26[29] 2016, 5271-5278. https://doi.org/10.1002/adfm.201601269

[101] Achira, H., Hoga, Y., Yoshikawa, I., Mutai, T., Matsumura, K., Houjou, H.,
       Polyhedron, 113, 2016, 123-131. https://doi.org/10.1016/j.poly.2016.04.022

[102] Geng, H., Luo, K., Zou, G., Wang, H., Ni, H., Yu, W., Li, Q., Wang, Y., New
       Journal of Chemistry, 40[12] 2016, 10371-10377.

[103] Han, A., Du, P., Sun, Z., Wu, H., Jia, H., Zhang, R., Liang, Z., Cao, R., Eisenberg,
       R., Inorganic Chemistry, 53[7] 2014, 3338-3344.

[104] Ni, J., Wang, Y.G., Wang, H.H., Xu, L., Zhao, Y.Q., Pan, Y.Z., Zhang, J.J., Dalton
       Transactions, 43[1] 2014, 352-360.

[105] Ni, J., Wang, Y.G., Wang, H.H., Pan, Y.Z., Xu, L., Zhao, Y.Q., Liu, X.Y., Zhang,

J.J., European Journal of Inorganic Chemistry, 6, 2014, 986-993.

[106] Shan, G.G., Li, H.B., Cao, H.T., Sun, H.Z., Zhu, D.X., Su, Z.M., Dyes and Pigments, 99[3] 2013, 1082-1090.

[107] Sun, H., Liu, S., Lin, W., Zhang, K.Y., Lv, W., Huang, X., Huo, F., Yang, H., Jenkins, G., Zhao, Q., Huang, W., Nature Communications, 5, 2014, 3601.

[108] Zhao, K.Y., Mao, H.T., Wen, L.L., Shan, G.G., Fu, Q., Sun, H.Z., Su, Z.M., Journal of Materials Chemistry C, 6[43] 2018, 11686-11693.

[109] Zhao, K.Y., Shan, G.G., Fu, Q., Su, Z.M., Organometallics, 35[23] 2016, 3996-4001.

[110] Hu, Y., Dong, Y., Sun, X., Zuo, G., Yin, J., Liu, S.H., Dyes and Pigments, 156, 2018, 260-266.

[111] Tsukuda, T., Kawase, M., Dairiki, A., Matsumoto, K., Tsubomura, T., Chemical Communications, 46, 2010, 1905.

[112] Babashkina, M., Safin, D., Bolte, M. Garcia, Y., Dalton Transactions, 40, 2011, 8523.

[113] Toma, O., Mercier, N., Botta, C., Journal of Materials Chemistry C, 4[25] 2016, 5940-5944.

[114] Toma, O., Allain, M., Meinardi, F., Forni, A., Botta, C., Mercier, N., Angewandte Chemie, 55[28] 2016, 7998-8002.

[115] Yang, Y., Yang, X., Fang, X., Wang, K.Z., Yan, D., Advanced Science, 5[11] 2018, 1801187.

[116] Wen, T., Zheng, Y., Xu, C., Zhang, J., Jaroniec, M., Qiao, S.Z., Materials Horizons, 5[6] 2018, 1151-1155.

[117] Yan, Y., Chen, J., Zhang, N.N., Wang, M.S., Sun, C., Xing, X.S., Li, R., Xu, J.G., Zheng, F.K., Guo, G.C., Dalton Transactions, 45[45] 2016, 18074-18078. https://doi.org/10.1039/c6dt03794a

[118] Mehlana, G., Wilkinson, C., Ramon, G., Bourne, S.A., Polyhedron, 98, 2015, 224-229.

[119] Sun, J.K., Chen, C., Cai, L.X., Ren, C.X., Tan, B., Zhang, J., Chemical Communications, 50[100] 2014, 15956-15959.

[120] Abedi, A., Safari, N., Amani, V., Khavasi, H.R., Dalton Transactions, 40[26] 2011, 6877-6885.

[121] Toma, O., Mercier, N., Allain, M., Meinardi, F., Botta, C., European Journal of Inorganic Chemistry, 4, 2017, 844-850.

[122] Song, X., Yu, H., Yan, X., Zhang, Y., Miao, Y., Ye, K., Wang, Y., Dalton Transactions, 47[17] 2018, 6146-6155.

[123] Shen, X.Y., Wang, Y.J., Zhao, E., Yuan, W.Z., Liu, Y., Lu, P., Qin, A., Ma, Y., Sun, J.Z., Tang, B.Z., Journal of Physical Chemistry C, 117[14] 2013, 7334-7347.

[124] Anuradha, La, D.D., Al Kobaisi, M., Bhosale, S.V., Scientific Reports, 5, 2015, 15652.

[125] Wang, Y., Xu, D., Gao, H., Wang, Y., Liu, X., Han, A., Zhang, C., Zang, L., Dyes and Pigments, 156, 2018, 291-298. https://doi.org/10.1016/j.dyepig.2018.04.021

[126] Yin, Y., Zhao, F., Chen, Z., Liu, G., Pu, S., Tetrahedron Letters, 59[50] 2018, 4416-4419.

[127] Ma, X., Hu, L., Han, X., Yin, J., Chinese Chemical Letters, 29[10] 2018, 1489-1492.

[128] Zhao, F., Chen, Z., Liu, G., Fan, C., Pu, S., Tetrahedron Letters, 59[9] 2018, 836-840.

[129] Zhao, S.S., Chen, L., Wang, L., Xie, Z., Chemical Communications, 53[52] 2017, 7048-7051.

[130] Li, Y., Zhuang, Z., Lin, G., Wang, Z., Shen, P., Xiong, Y., Wang, B., Chen, S., Zhao, Z., Tang, B.Z., New Journal of Chemistry, 42[6] 2018, 4089-4094.

[131] Chen, J.R., Zhao, J., Xu, B.J., Yang, Z.Y., Liu, S.W., Xu, J.R., Zhang, Y., Wu, Y.C., Lv, P.Y., Chi, Z.G., Chinese Journal of Polymer Science, 35[2] 2017, 282-292.

[132] Gundu, S., Kim, M., Mergu, N., Son, Y.A., Dyes and Pigments, 146, 2017, 7-13.

[133] Umar, S., Jha, A.K., Purohit, D., Goel, A., Journal of Organic Chemistry, 82[9] 2017, 4766-4773.

[134] Jadhav, T., Dhokale, B., Patil, Y., Misra, R., RSC Advances, 5[83] 2015, 68187-68191. https://doi.org/10.1039/c5ra12697e

[135] Jadhav, T., Dhokale, B., Misra, R., Journal of Materials Chemistry C, 3[35] 2015, 9063-9068.

[136] Misra, R., Jadhav, T., Dhokale, B., Mobin, S.M., Chemical Communications, 50[65] 2014, 9076-9078.

[137] Hu, T., Yao, B., Chen, X., Li, W., Song, Z., Qin, A., Sun, J.Z., Tang, B.Z., Chemical Communications, 51[42] 2015, 8849-8852.

[138] Tong, J., Wang, Y., Mei, J., Wang, J., Qin, A., Sun, J.Z., Tang, B.Z., Chemistry - a European Journal, 20[16] 2014, 4661-4670.

[139] Qi, Y., Liu, W., Wang, Y., Ma, L., Yu, Y., Zhang, Y., Ren, L., New Journal of Chemistry, 42[14] 2018, 11373-11380.

[140] Qi, Y., Wang, Y., Yu, Y., Liu, Z., Zhang, Y., Du, G., Qi, Y., RSC Advances, 6[40] 2016, 33755-33762.

[141] Salimimarand, M., La, D.D., Bhosale, S.V., Jones, L.A., Bhosale, S.V., Applied Sciences, 7[11] 2017, 1119.

[142] Salimimarand, M., La, D.D., Kobaisi, M.A., Bhosale, S.V., Scientific Reports, 7, 2017, 42898.

[143] La, D.D., Anuradha, A., Hundal, A.K., Bhosale, S.V., Jones, L.A., Bhosale, S.V., Supramolecular Chemistry, 30[1] 2018, 1-8.

[144] Qi, Y., Wang, Y., Ge, G., Liu, Z., Yu, Y., Xue, M., Journal of Materials Chemistry C, 5[42] 2017, 11030-11038.

[145] Liu, X., Li, M., Liu, M., Yang, Q., Chen, Y., Chemistry - a European Journal, 24[50] 2018, 13197-13204.

[146] Pan, L., Cai, Y., Wu, H., Zhou, F., Qin, A., Wang, Z., Tang, B.Z., Materials Chemistry Frontiers, 2[7] 2018, 1310-1316.

[147] Ruan, Z., Li, L., Wang, C., Xie, Y., Hu, Q., Peng, Q., Ye, S., Li, Q., Li, Z., Small, 12[47] 2016, 6623-6632.

[148] Chen, M., Chen, R., Shi, Y., Wang, J., Cheng, Y., Li, Y., Gao, X., Yan, Y., Sun, J.Z., Qin, A., Kwok, R.T.K., Lam, J.W.Y., Tang, B.Z., Advanced Functional Materials, 28[6] 2018, 1704689. https://doi.org/10.1002/adfm.201704689

[149] Ma, Z., Wang, Z., Meng, X., Ma, Z., Xu, Z., Ma, Y., Jia, X., Angewandte Chemie, 55[2] 2016, 519-522.

[150] Jadhav, T., Dhokale, B., Mobin, S.M., Misra, R., RSC Advances, 5[38] 2015, 29878-29884. https://doi.org/10.1039/c5ra04881h

[151] Zhang, Y., Mao, H., Kong, L., Tian, Y., Tian, Z., Zeng, X., Zhi, J., Shi, J., Tong, B., Dong, Y., Dyes and Pigments, 133, 2016, 354-362. https://doi.org/10.1016/j.dyepig.2016.06.016

[152] Sun, J., Zhang, G., Jia, X., Xue, P., Jia, J., Lu, R., Acta Chimica Sinica, 74[2] 2016, 165-171.

[153] Ekbote, A., Mobin, S.M., Misra, R., Journal of Materials Chemistry C, 6[40] 2018, 10888-10901.

[154] Teng, M.J., Jia, X.R., Chen, X.F., Wei, Y., Angewandte Chemie, 51[26] 2012, 6398-6401.

[155] Ma, Z., Teng, M., Wang, Z., Yang, S., Jia, X., Angewandte Chemie, 52[47] 2013, 12268-12272.

[156] Ma, Z., Teng, M., Wang, Z., Jia, X., Tetrahedron Letters, 54[48] 2013, 6504-6506.

[157] Han, X., Liu, Y., Liu, G., Luo, J., Liu, S.H., Zhao, W., Yin, J., Chemistry - an Asian Journal, 14[6] 2019, 890-895.

[158] Zhang, H., Nie, Y., Miao, J., Zhang, D., Li, Y., Liu, G., Sun, G., Jiang, X., Journal of Materials Chemistry C, 7[11] 2019, 3306-3314.

[159] Huang, G., Jiang, Y., Yang, S., Li, B.S., Tang, B.Z., Advanced Functional Materials, 2019, 1900516.

[160] Chen, P.Z., Wang, J.X., Niu, L.Y., Chen, Y.Z., Yang, Q.Z., Journal of Materials Chemistry C, 5[47] 2017, 12538-12546.

[161] Butler, T., Zhuang, M., Fraser, C.L., Journal of Physical Chemistry C, 122[33] 2018, 19090-19099.

[162] Morris, W.A., Butler, T., Kolpaczynska, M., Fraser, C.L., Materials Chemistry Frontiers, 1[1] 2017, 158-166.

[163] Morris, W.A., Kolpaczynska, M., Fraser, C.L., Journal of Physical Chemistry C, 120[39] 2016, 22539-22548.

[164] Morris, W.A., Sabat, M., Butler, T., Derosa, C.A., Fraser, C.L., Journal of Physical Chemistry C, 120[26] 2016, 14289-14300.

[165] Butler, T., Morris, W.A., Samonina-Kosicka, J., Fraser, C.L., ACS Applied Materials and Interfaces, 8[2] 2016, 1242-1251.

[166] Butler, T., Morris, W.A., Samonina-Kosicka, J., Fraser, C.L., Chemical Communications, 51[16] 2015, 3359-3362. https://doi.org/10.1039/c4cc09439e

[167] Morris, W.A., Liu, T., Fraser, C.L., Journal of Materials Chemistry C, 3[2] 2015, 352-363.

[168] Liu, T., Chien, A.D., Lu, J., Zhang, G., Fraser, C.L., Journal of Materials

Chemistry, 21[23] 2011, 8401-8408.

[169] Nguyen, N.D., Zhang, G., Lu, J., Sherman, A.E., Fraser, C.L., Journal of Materials Chemistry, 21[23] 2011, 8409-8415.

[170] Zhang, G., Singer, J.P., Kooi, S.E., Evans, R.E., Thomas, E.L., Fraser, C.L., Journal of Materials Chemistry, 21[23] 2011, 8295-8299.

[171] Louis, M., Brosseau, A., Guillot, R., Ito, F., Allain, C., Métivier, R., Journal of Physical Chemistry C, 121[29] 2017, 15897-15907. https://doi.org/10.1021/acs.jpcc.7b01901

[172] Butler, T., Mathew, A.S., Sabat, M., Fraser, C.L., ACS Applied Materials and Interfaces, 9[20] 2017, 17603-17612.

[173] Butler, T., Wang, F., Sabat, M., Fraser, C.L., Materials Chemistry Frontiers, 1[9] 2017, 1804-1817.

[174] Tan, R., Lin, Q., Wen, Y., Xiao, S., Wang, S., Zhang, R., Yi, T., CrystEngComm, 17[35] 2015, 6674-6680. https://doi.org/10.1039/c5ce01138h

[175] Li, M., Han, Y., Zhang, Z., He, X., Chen, Y., Dyes and Pigments, 166, 2019, 159-167.

[176] Xie, Z., Su, T., Ubba, E., Deng, H., Mao, Z., Yu, T., Zheng, T., Zhang, Y., Liu, S., Chi, Z., Journal of Materials Chemistry C, 7[11] 2019, 3300-3305.

[177] Hsu, L.Y., Maity, S., Matsunaga, Y., Hsu, Y.F., Liu, Y.H., Peng, S.M., Shinmyozu, T., Yang, J.S., Chemical Science, 9[48] 2018, 8990-9001.

[178] Li, A., Liu, Y., Han, L., Xu, S., Song, C., Geng, Y., Pan, L., Xu, B., Tian, W., Zhang, H., Xu, W., Cui, H., Dyes and Pigments, 161, 2019, 182-187. https://doi.org/10.1016/j.dyepig.2018.09.041

[179] Li, A., Ma, Z., Wu, J., Li, P., Wang, H., Geng, Y., Xu, S., Yang, B., Zhang, H., Cui, H., Xu, W., Advanced Optical Materials, 6[3] 2018, 1700647. https://doi.org/10.1002/adom.201700647

[180] Sun, Q., Wang, H., Xu, X., Lu, Y., Xue, S., Zhang, H., Yang, W., Dyes and Pigments, 149, 2018, 407-414.

[181] He, Z., Zhang, L., Mei, J., Zhang, T., Lam, J.W.Y., Shuai, Z., Dong, Y.Q., Tang, B.Z., Chemistry of Materials, 27[19] 2015, 6601-6607.

[182] Ma, Z., Ji, Y., Lan, Y., Kuang, G.C., Jia, X., Journal of Materials Chemistry C, 6[9] 2018, 2270-2274.

[183] Ma, Z., Wang, Z., Li, Y., Song, S., Jia, X., Tetrahedron Letters, 57[48] 2016, 5377-5380.

[184] Ma, Z., Yang, F., Wang, Z., Jia, X., Tetrahedron Letters, 56[2] 2015, 393-396.

[185] Dai, Q., Zhang, J., Tan, R., Wang, S., Li, Y., Li, Q., Xiao, S., Materials Letters, 164, 2016, 239-242.

[186] Tu, D., Leong, P., Li, Z., Hu, R., Shi, C., Zhang, K.Y., Yan, H., Zhao, Q., Chemical Communications, 52[84] 2016, 2494-12497.

[187] Xue, P., Yao, B., Liu, X., Sun, J., Gong, P., Zhang, Z., Qian, C., Zhang, Y., Lu, R., Journal of Materials Chemistry C, 3[5] 2015, 1018-1025.

[188] Sun, Q., Qiu, X., Lu, Y., Xu, X., Wang, H., Xue, S., Yang, W., Journal of Materials Chemistry C, 5[35] 2017, 9157-9164.

[189] Teng, M., Wang, Z., Ma, Z., Chen, X., Jia, X., RSC Advances, 4[39] 2014, 20239-20241.

[190] Matsumoto, S., Moteki, J., Ito, Y., Akazome, M., Tetrahedron Letters, 58[36] 2017, 3512-3516.

[191] He, G., Du, L., Gong, Y., Liu, Y., Yu, C., Wei, C., Yuan, W.Z., ACS Omega, 4[1] 2019, 344-351.

[192] Wang, Z.Y., Zhao, J.W., Li, P., Feng, T., Wang, W.J., Tao, S.L., Tong, Q.X., New Journal of Chemistry, 42[11] 2018, 8924-8932.

[193] Ito, S., Taguchi, T., Yamada, T., Ubukata, T., Yamaguchi, Y., Asami, M., RSC Advances, 7[28] 2017, 16953-16962. Photographs abstracted from this publication under Creative Commons Licence 3.0. https://doi.org/10.1039/c7ra01006k

[194] Horak, E., Robić, M., Šimanović, A., Mandić, V., Vianello, R., Hranjec, M., Steinberg, I.M., Dyes and Pigments, 162, 2019, 688-696. https://doi.org/10.1016/j.dyepig.2018.10.069

[195] Zhu, H., Chen, P., Kong, L., Tian, Y., Yang, J., Journal of Physical Chemistry C, 122[34] 2018, 19793-19800.

[196] Chen, J., Li, D., Chi, W., Liu, G., Liu, S.H., Liu, X., Zhang, C., Yin, J., Chemistry - A European Journal, 24[15] 2018, 3671-3676.

[197] Ito, S., Yamada, T., Taguchi, T., Yamaguchi, Y., Asami, M., Chemistry - an Asian Journal, 11[13] 2016, 1963-1970.

[198] Guo, C., Li, M., Yuan, W., Wang, K., Zou, B., Chen, Y., Journal of Physical

Chemistry C, 121[48] 2017, 27009-27017.

[199] Wang, S., Tan, R., Li, Y., Li, Q., Xiao, S., Dyes and Pigments, 132, 2016, 342-346.

[200] Jadhav, T., Choi, J.M., Shinde, J., Lee, J.Y., Misra, R., Journal of Materials Chemistry C, 5[24] 2017, 6014-6020.

[201] Jadhav, T., Choi, J.M., Dhokale, B., Mobin, S.M., Lee, J.Y., Misra, R., Journal of Physical Chemistry C, 120[33] 2016, 18487-18495. https://doi.org/10.1021/acs.jpcc.6b06277

[202] Li, G., Zhao, J., Zhang, D., Shi, Z., Zhu, Z., Song, H., Zhu, J., Tao, S., Lu, F., Tong, Q., Journal of Materials Chemistry C, 4[37] 2016, 8787-8794.

[203] Jadhav, T., Dhokale, B., Mobin, S.M., Misra, R., Journal of Materials Chemistry C, 3[38] 2015, 9981-9988.

[204] Zhan, Y., Gong, P., Yang, P., Jin, Z., Bao, Y., Li, Y., Xu, Y., RSC Advances, 6[39] 2016, 32697-32704. https://doi.org/10.1039/c6ra03310e

[205] Zhao, F., Chen, Z., Fan, C., Liu, G., Pu, S., Dyes and Pigments, 164, 2019, 390-397.

[206] Zhan, Y., Zhao, J., Yang, P., Ye, W., RSC Advances, 6[94] 2016, 92144-92151. https://doi.org/10.1039/c6ra19791d

[207] Xue, P., Yao, B., Sun, J., Xu, Q., Chen, P., Zhang, Z., Lu, R., Journal of Materials Chemistry C, 2[20] 2014, 3942-3950.

[208] Yan, D., Yang, H., Meng, Q., Lin, H., Wei, M., Advanced Functional Materials, 24[5] 2014, 587-594.

[209] Zhu, Y.Y., Xia, H.Y., Yao, L.F., Huang, D.P., Song, J.Y., He, H.F., Shen, L., Zhao, F., RSC Advances, 9[13] 2019, 7176-7180.

[210] Macchione, M., Tsemperouli, M., Goujon, A., Mallia, A.R., Sakai, N., Sugihara, K., Matile, S., Helvetica Chimica Acta, 101[4] 2018, e1800014. https://doi.org/10.1002/hlca.201800014

[211] Mitani, M., Ogata, S., Yamane, S., Yoshio, M., Hasegawa, M., Kato, T., Journal of Materials Chemistry C, 4[14] 2016, 2752-2760.

[212] Gan, K.P., Yoshio, M., Kato, T., Journal of Materials Chemistry C, 4[22] 2016, 5073-5080.

[213] Mitani, M., Yamane, S., Yoshio, M., Funahashi, M., Kato, T., Molecular Crystals

and Liquid Crystals, 594[1] 2014, 112-121.
https://doi.org/10.1080/15421406.2014.917499

[214] Mosca, S., Milani, A., Peña-Álvarez, M., Yamaguchi, S., Hernández, V., Ruiz Delgado, M.C., Castiglioni, C., Journal of Physical Chemistry C, 122[30] 2018, 17537-17543. https://doi.org/10.1021/acs.jpcc.8b05423

[215] Nagura, K., Saito, S., Yusa, H., Yamawaki, H., Fujihisa, H., Sato, H., Shimoikeda, Y., Yamaguchi, S., Journal of the American Chemical Society, 135[28] 2013, 10322-10325. https://doi.org/10.1021/ja4055228

[216] Nallicheri, R.A., Rubner, M.F., Macromolecules, 24[2] 1991, 517-525.

[217] Kim, S.J., Reneker, D.H., Polymer Bulletin, 31[3] 1993, 367-374.

[218] Crenshaw, B.R., Weder, C., Macromolecules, 39[26] 2006, 9581-9589.

[219] Harvey, C.P., Tovar, J.D., Journal of Polymer Science, Part A: Polymer Chemistry, 49[22] 2011, 4861-4874.

[220] Bao, S., Li, J., Lee, K.I., Shao, S., Hao, J., Fei, B., Xin, J.H., ACS Applied Materials and Interfaces, 5[11] 2013, 4625-4631. https://doi.org/10.1021/am4013648

[221] Cellini, F., Khapli, S., Peterson, S.D., Porfiri, M., Applied Physics Letters, 105[6] 2014, 061907. https://doi.org/10.1063/1.4893010

[222] Cellini, F., Zhou, L., Khapli, S., Peterson, S.D., Porfiri, M., Mechanics of Materials, 93, 2016, 145-162. https://doi.org/10.1016/j.mechmat.2015.10.013

[223] Wu, Y., Hu, J., Huang, H., Li, J., Zhu, Y., Tang, B., Han, J., Li, L., Journal of Polymer Science B, 52[2] 2014, 104-110.

[224] Imato, K., Kanehara, T., Ohishi, T., Nishihara, M., Yajima, H., Ito, M., Takahara, A., Otsuka, H., ACS Macro Letters, 4[11] 2015, 1307-1311. https://doi.org/10.1021/acsmacrolett.5b00717

[225] Wang, Z., Ma, Z., Wang, Y., Xu, Z., Luo, Y., Wei, Y., Jia, X., Advanced Materials, 27[41] 2015, 6469-6474.

[226] Sagara, Y., Karman, M., Verde-Sesto, E., Matsuo, K., Kim, Y., Tamaoki, N., Weder, C., Journal of the American Chemical Society, 140[5] 2018, 1584-1587. https://doi.org/10.1021/jacs.7b12405

[227] Wang, L., Zhou, W., Tang, Q., Yang, H., Zhou, Q., Zhang, X., Polymers, 10[9] 2018, 994.

[228] Cellini, F., Block, L., Li, J., Khapli, S., Peterson, S.D., Porfiri, M., Sensors and Actuators, B: Chemical, 234, 2016, 510-520. https://doi.org/10.1016/j.snb.2016.04.149

[229] Yue, Y., Kurokawa, T., Haque, M.A., Nakajima, T., Nonoyama, T., Li, X., Kajiwara, I., Gong, J.P., Nature Communications, 5, 2014, 4659.

[230] Wang, X.Q., Hong, R., Wang, C.F., Tan, P.F., Ji, W.Q., Chen, S., Materials Letters, 189, 2017, 321-324.

[231] Jia, X., Wang, J., Wang, K., Zhu, J., Langmuir, 31[31] 2015, 8732-8737.

[232] Hong, R., Shi, Y., Wang, X.Q., Peng, L., Wu, X., Cheng, H., Chen, S., RSC Advances, 7[53] 2017, 33258-33262. https://doi.org/10.1039/c7ra05622b

[233] Cho, Y., Lee, S.Y., Ellerthorpe, L., Feng, G., Lin, G., Wu, G., Yin, J., Yang, S., Advanced Functional Materials, 25[38] 2015, 6041-6049.

[234] Ying, Y., Xia, J., Foulger, S.H., Applied Physics Letters, 90[7] 2007, 071110.

[235] Zhu, Q., Van Vliet, K., Holten-Andersen, N., Miserez, A., Advanced Functional Materials, 29[14] 2019, 1808191. https://doi.org/10.1002/adfm.201808191

[236] Zhang, G., Sun, J., Xue, P., Zhang, Z., Gong, P., Peng, J., Lu, R., Journal of Materials Chemistry C, 3[12] 2015, 2925-2932.

[237] Zhang, Y., Wang, K., Zhuang, G., Xie, Z., Zhang, C., Cao, F., Pan, G., Chen, H., Zou, B., Ma, Y., Chemistry - A European Journal, 21[6] 2015, 2474-2479.

[238] Gong, Y., Zhang, Y., Yuan, W.Z., Sun, J.Z., Zhang, Y., Journal of Physical Chemistry C, 118[20] 2014, 10998-11005.

[239] Yoon, S.J., Park, S., Journal of Materials Chemistry, 21[23] 2011, 8338-8346.

[240] Adak, A., Panda, T., Raveendran, A., Bejoymohandas, K.S., Asha, K.S., Prakasham, A.P., Mukhopadhyay, B., Panda, M.K., ACS Omega, 3[5] 2018, 5291-5300. https://doi.org/10.1021/acsomega.8b00250

[241] Wang, B., Wei, C., RSC Advances, 8[40] 2018, 22806-22812.

[242] Li, S., Sun, J., Qile, M., Cao, F., Zhang, Y., Song, Q., ChemPhysChem, 18[11] 2017, 1481-1485.

[243] Shi, P., Duan, Y., Wei, W., Xu, Z., Li, Z., Han, T., Journal of Materials Chemistry C, 6[10] 2018, 2476-2482.

[244] Han, T., Gu, X., Lam, J.W.Y., Leung, A.C.S., Kwok, R.T.K., Han, T., Tong, B., Shi, J., Dong, Y., Tang, B.Z., Journal of Materials Chemistry C, 4[44] 2016,

10430-10434. https://doi.org/10.1039/c6tc03883b

[245] Tian, H., Tang, X., Dong, Y.Q., Molecules, 22[12] 2017, 2222.

[246] Zhu, H., Weng, S., Zhang, H., Yu, H., Kong, L., Zhong, Y., Tian, Y., Yang, J., CrystEngComm, 20[20] 2018, 2772-2779.

[247] Hariharan, P.S., Prasad, V.K., Nandi, S., Anoop, A., Moon, D., Anthony, S.P., Crystal Growth and Design, 17[1] 2017, 146-155.

[248] Zhang, Y., Feng, Y.Q., Wang, J.H., Han, G., Li, M.Y., Xiao, Y., Feng, Z.D., RSC Advances, 7[57] 2017, 35672-35680.

[249] Jia, J., Zhao, H., Tetrahedron Letters, 60[3] 2019, 252-259.

[250] Zhang, Y., Zhuang, G., Ouyang, M., Hu, B., Song, Q., Sun, J., Zhang, C., Gu, C., Xu, Y., Ma, Y., Dyes and Pigments, 98[3] 2013, 486-492. https://doi.org/10.1016/j.dyepig.2013.03.017

[251] Sun, J., Lv, X., Wang, P., Zhang, Y., Dai, Y., Wu, Q., Ouyang, M., Zhang, C., Journal of Materials Chemistry C, 2[27] 2014, 5365-5371.

[252] Sun, J., Dai, Y., Ouyang, M., Zhang, Y., Zhan, L., Zhang, C., Journal of Materials Chemistry C, 3, 2015, 3356-3363.

[253] Zhang, J., Chen, Z., Yang, L., Hu, F., Yu, G.A., Yin, J., Liu, S.H., Dyes and Pigments, 136, 2017, 168-174.

[254] Hariharan, P.S., Mothi, E.M., Moon, D., Anthony, S.P., ACS Applied Materials and Interfaces, 8[48] 2016, 33034-33042. https://doi.org/10.1021/acsami.6b11939

[255] Ishi-I, T., Tanaka, H., Youfu, R., Aizawa, N., Yasuda, T., Kato, S.I., Matsumoto, T., New Journal of Chemistry, 43[13] 2019, 4998-5010.

[256] Yu, X., Ge, X., Lan, H., Li, Y., Geng, L., Zhen, X., Yi, T., ACS Applied Materials and Interfaces, 7[43] 2015, 24312-24321. https://doi.org/10.1021/acsami.5b08402

[257] Yu, X., Xie, D., Lan, H., Li, Y., Zhen, X., Ren, J., Yi, T., Journal of Materials Chemistry C, 5[24] 2017, 5910-5916.

[258] Peebles, C., Wight, C.D., Iverson, B.L., Journal of Materials Chemistry C, 3[46] 2015, 12156-12163.

[259] Srivastava, A.K., Singh, A.K., Kumari, N., Yadav, R., Gulino, A., Speghini, A., Nagarajan, R., Mishra, L., Journal of Luminescence, 182, 2017, 274-282. https://doi.org/10.1016/j.jlumin.2016.10.042

[260] Ma, Z., Wang, Z., Xu, Z., Jia, X., Wei, Y., Journal of Materials Chemistry C,

3[14] 2015, 3399-3405.

[261] Sagara, Y., Komatsu, T., Ueno, T., Hanaoka, K., Kato, T., Nagano, T., Advanced Functional Materials, 23[42] 2013, 5277-5284.

[262] Nagai, A., Okabe, Y., Chemical Communications, 50[70] 2014, 10052-10054.

[263] Kong, Q., Zhuang, W., Li, G., Xu, Y., Jiang, Q., Wang, Y., New Journal of Chemistry, 41[22] 2017, 13784-13791.

[264] Li, W., Wang, L., Zhang, J.P., Wang, H., Journal of Materials Chemistry C, 2[10] 2014, 1887-1892.

[265] Ishizuki, K., Oka, H., Aoki, D., Goseki, R., Otsuka, H., Chemistry - a European Journal, 24[13] 2018, 3170-3173.

[266] Koevoets, R.A., Karthikeyan, S., Magusin, P.C.M.M., Meijer, E.W., Sijbesma, R.P., Macromolecules, 42[7] 2009, 2609-2617. https://doi.org/10.1021/ma801220q

[267] Kondo, M., Nakanishi, T., Matsushita, T., Kawatsuki, N., Macromolecular Chemistry and Physics, 218[1] 2017, 1600321.

[268] Kosuge, T., Imato, K., Goseki, R., Otsuka, H., Macromolecules, 49[16] 2016, 5903-5911. https://doi.org/10.1021/acs.macromol.6b01333

[269] Williams, G.A., Ishige, R., Cromwell, O.R., Chung, J., Takahara, A., Guan, Z., Advanced Materials, 27[26] 2015, 3934-3941.

[270] Ruiz De Luzuriaga, A., Matxain, J.M., Ruipérez, F., Martin, R., Asua, J.M., Cabañero, G., Odriozola, I., Journal of Materials Chemistry C, 4[26] 2016, 6220-6223. https://doi.org/10.1039/c6tc02383e

[271] Toivola, R., Jang, S.H., Mannikko, D., Stoll, S., Jen, A.K.Y., Flinn, B.D., Polymer, 142, 2018, 132-143. https://doi.org/10.1016/j.polymer.2018.03.029

[272] Toivola, R., Lai, P.N., Yang, J., Jang, S.H., Jen, A.K.Y., Flinn, B.D., Composites Science and Technology, 139, 2017, 74-82.

[273] Li, G., Xu, Y., Kong, Q., Zhuang, W., Wang, Y., Journal of Materials Chemistry C, 5[33] 2017, 8527-8534.

[274] Li, G., Xu, Y., Zhuang, W., Wang, Y., RSC Advances, 6[88] 2016, 84787-84793.

[275] Zhou, Y., Qian, L., Liu, M., Wu, G., Gao, W., Ding, J., Huang, X., Wu, H., RSC Advances, 7[81] 2017, 51444-51451. Photographs abstracted from this publication under Creative Commons Licence 3.0. https://doi.org/10.1039/c7ra09515e

[276] Li, A., Li, P., Geng, Y., Xu, S., Zhang, H., Cui, H., Xu, W., Spectrochimica Acta A, 202, 2018, 70-75.

[277] Lei, Y., Zhou, Y., Qian, L., Wang, Y., Liu, M., Huang, X., Wu, G., Wu, H., Ding, J., Cheng, Y., Journal of Materials Chemistry C, 5[21] 2017, 5183-5192.

[278] Kondo, M., Yamoto, T., Miura, S., Hashimoto, M., Kitamura, C., Kawatsuki, N., Chemistry - an Asian Journal, 14[3] 2019, 471-479.

[279] Wu, J., Cheng, Y., Lan, J., Wu, D., Qian, S., Yan, L., He, Z., Li, X., Wang, K., Zou, B., You, J., Journal of the American Chemical Society, 138[39] 2016, 12803-12812. https://doi.org/10.1021/jacs.6b03890

[280] Qi, F., Lin, J., Wang, X., Cui, P., Yan, H., Gong, S., Ma, C., Liu, Z., Huang, W., Dalton Transactions, 45[17] 2016, 7278-7284.

[281] Okazaki, M., Takeda, Y., Data, P., Pander, P., Higginbotham, H., Monkman, A.P., Minakata, S., Chemical Science, 8[4] 2017, 2677-2686.

[282] Takeda, Y., Kaihara, T. Okazaki, M., Higginbotham, H., Data, P., Tohnai, N., Minakata, S., Chemical Communications, 54, 2018, 6847-6850. Photographs abstracted from this reference (DOI: 10.1039/c8cc02365d) under Creative Commons Licence 3.0.

[283] Zhao, H., Wang, Y., Harrington, S., Ma, L., Hu, S., Wu, X., Tang, H., Xue, M., Wang, Y., RSC Advances, 6[71] 2016, 66477-66483. https://doi.org/10.1039/c6ra14707k

[284] Fang, W., Zhang, Y., Zhang, G., Kong, L., Yang, L., Yang, J., CrystEngComm, 19[9] 2017, 1294-1303.

[285] Ying, S., Chen, M., Liu, Z., Zhang, K., Pan, Y., Xue, S., Yang, W., Journal of Materials Chemistry C, 4[34] 2016, 8006-8013.

[286] Zhou, Y., Qian, L., Liu, M., Huang, X., Wang, Y., Cheng, Y., Gao, W., Wu, G., Wu, H., Journal of Materials Chemistry C, 5[36] 2017, 9264-9272.

[287] Yang, W., Liu, C., Lu, S., Du, J., Gao, Q., Zhang, R., Liu, Y., Yang, C., Journal of Materials Chemistry C, 6[2] 2018, 290-298.

[288] Tanioka, M., Kamino, S., Muranaka, A., Ooyama, Y., Ota, H., Shirasaki, Y., Horigome, J., Ueda, M., Uchiyama, M., Sawada, D., Enomoto, S., Journal of the American Chemical Society, 137[20] 2015, 6436-6439. https://doi.org/10.1021/jacs.5b00877

[289] Kamino, S., Tanioka, M., Sawada, D., Journal of Synthetic Organic Chemistry,

76[10] 2018, 1066-1075.

[290] Miyagi, K., Teramoto, Y., Journal of Materials Chemistry C, 6[6] 2018, 1370-1376.

[291] Suzuki, T., Okada, H., Nakagawa, T., Komatsu, K., Fujimoto, C., Kagi, H., Matsuo, Y., Chemical Science, 9[2] 2018, 475-482.

[292] Roy, B., Reddy, M.C., Hazra, P., Chemical Science, 9[14] 2018, 3592-3606.

[293] Duan, Y., Xiang, X., Dong, Y., Acta Chimica Sinica, 74[11] 2016, 923-928.

[294] Krishna, G.R., Devarapalli, R., Prusty, R., Liu, T., Fraser, C.L., Ramamurty, U., Reddy, C.M., IUCrJ, 2, 2015, 611-619.

[295] Hou, Y., Li, Z., Hou, J., Shi, P., Li, Y., Niu, M., Liu, Y., Han, T., Dyes and Pigments, 159, 2018, 252-261.

[296] Hu, J., Wang, J., Liu, R., Li, Y., Lu, J., Zhu, H., Dyes and Pigments, 153, 2018, 84-91.

[297] Zhang, K., Chen, M., Liu, Z., Ying, S., Zhang, H., Xue, S., Yang, W., Journal of Luminescence, 194, 2018, 588-593.

[298] Liu, J.J., Yang, J., Wang, J.L., Chang, Z.F., Li, B., Song, W.T., Zhao, Z., Lou, X., Dai, J., Xia, F., Materials Chemistry Frontiers, 2[6] 2018, 1126-1136.

[299] Xue, P., Yang, Z., Chen, P., Journal of Materials Chemistry C, 6[18] 2018, 4994-5000.

[300] Jia, J., Wu, Y., Dyes and Pigments, 147, 2017, 537-543.

[301] Sun, J., Han, J., Liu, Y., Duan, Y., Han, T., Yuan, J., Journal of Materials Chemistry C, 4[35] 2016, 8276-8283.

[302] Zhao, Z., Chen, T., Jiang, S., Liu, Z., Fang, D., Dong, Y.Q., Journal of Materials Chemistry C, 4[21] 2016, 4800-4804.

[303] Lv, Y., Liu, Y., Guo, D., Ye, X., Liu, G., Tao, X., Chemistry - an Asian Journal, 9[10] 2014, 2885-2890.

[304] Ito, S., Yamada, T., Asami, M., ChemPlusChem, 81[12] 2016, 1272-1275.

[305] Zhang, G., Lu, J., Sabat, M., Fraser, C.L., Journal of the American Chemical Society, 132[7] 2010, 2160-2162.

[306] Krishna, G.R., Kiran, M.S.R.N., Fraser, C.L., Ramamurty, U., Reddy, C.M., Advanced Functional Materials, 23[11] 2013, 1422-1430.

[307] Mizuguchi, K., Kageyama, H., Nakano, H., Materials Letters, 65[17-18] 2011, 2658-2661. https://doi.org/10.1016/j.matlet.2011.05.068

[308] Yamane, S., Sagara, Y., Mutai, T., Araki, K., Kato, T., Journal of Materials Chemistry C, 1[15] 2013, 2648-2656.

[309] Wei, J., Liang, B., Cheng, X., Zhang, Z., Zhang, H., Wang, Y., RSC Advances, 5[88] 2015, 71903-71910. https://doi.org/10.1039/c5ra12050k

[310] Hou, J., Wu, X., Sun, W., Duan, Y., Liu, Y., Han, T., Li, Z., Spectrochimica Acta A, 214, 2019, 348-354.

[311] Sagara, Y., Kubo, K., Nakamura, T., Tamaoki, N., Weder, C., Chemistry of Materials, 29[3] 2017, 1273-1278.

[312] Cao, Y., Xi, Y., Teng, X., Li, Y., Yan, X., Chen, L., Dyes and Pigments, 137, 2017, 75-83.

[313] Liu, D., Cao, Y., Yan, X., Wang, B., Research on Chemical Intermediates, 45[4] 2019, 2429-2439.

[314] Elbert, S.M., Wagner, P., Kanagasundaram, T., Rominger, F., Mastalerz, M., Chemistry - a European Journal, 23[4] 2017, 935-945. https://doi.org/10.1002/chem.201604421

[315] Oda, K., Hiroto, S., Shinokubo, H., Journal of Materials Chemistry C, 5[22] 2017, 5310-5315.

[316] Roy, S., Hazra, A., Bandyopadhyay, A., Raut, D., Madhuri, P.L., Rao, D.S.S., Ramamurty, U., Pati, S.K., Krishna Prasad, S., Maji, T.K., Journal of Physical Chemistry Letters, 7[20] 2016, 4086-4092. https://doi.org/10.1021/acs.jpclett.6b01891

[317] Lavrenova, A., Holtz, A., Simon, Y.C., Weder, C., Macromolecular Materials and Engineering, 301[5] 2016, 549-554.

[318] Han, J., Sun, J., Li, Y., Duan, Y., Han, T., Journal of Materials Chemistry C, 4[39] 2016, 9287-9293.

[319] Imoto, H., Kizaki, K., Naka, K., Chemistry - an Asian Journal, 10[8] 2015, 1698-1702.

[320] Mao, Z., Yang, Z., Mu, Y., Zhang, Y., Wang, Y.F., Chi, Z., Lo, C.C., Liu, S., Lien, A., Xu, J., Angewandte Chemie, 54[21] 2015, 6270-6273.

[321] Zhao, H., Wang, Y., Wang, Y., He, G., Xue, M., Guo, P., Dai, B., Liu, Z., Qi, Y.,

RSC Advances, 5[25] 2015, 19176-19181. https://doi.org/10.1039/c4ra16069j

[322] Zhang, Y., Han, T., Gu, S., Zhou, T., Zhao, C., Guo, Y., Feng, X., Tong, B., Bing, J., Shi, J., Zhi, J., Dong, Y., Chemistry - a European Journal, 20[29] 2014, 8856-8861.

[323] Qi, Q., Zhang, J., Xu, B., Li, B., Zhang, S.X.A., Tian, W., Journal of Physical Chemistry C, 117[47] 2013, 24997-25003.

[324] Sun, X., Zhang, X., Li, X., Liu, S., Zhang, G., Journal of Materials Chemistry, 22[33] 2012, 17332-17339.

[325] Roy, B., Reddy, M.C., Panja, S.N., Hazra, P., Journal of Physical Chemistry C, 123[6] 2019, 3848-3854.

Keyword Index

Materials Research Forum LLC
https://doi.org/10.21741/9781644900277

www.ingramcontent.com/pod-product-compliance
Lightning Source LLC
Chambersburg PA
CBHW071657210326
41597CB00017B/2232